How to Become a

First-Generation Farmer

by **John Terry**

Hon. B.A., Cert.Ed., C.Biol., M.S.B.,
M.C.I.Hort., F.R.Ag.S., F.R.S.A.

How to Become a

First-Generation Farmer

by **John Terry**

Hon. B.A., Cert.Ed., C.Biol., M.S.B.,
M.C.I.Hort., F.R.Ag.S., F.R.S.A.

By the same author:

Pigs in the Playground
Calves in the Classroom
Ducks in Detention
Rabbits on Report

Published by 5m Publishing
Benchmark House
8 Smithy Wood Drive
Sheffield
S35 1QN
United Kingdom

www.5mpublishing.com
books@5mpublishing.com

Cover images: John Terry

ISBN: 978-1-908397-928

A CIP catalogue record for this book is available from the British Library.

Typeset by David Exley www.beamreachuk.co.uk

Printed by www.beamreachuk.co.uk

For my lovely wife, Sarah, and our two children, Jonathan
and Roseanna - not forgetting my dear mother and
father, and also my Uncle Ben and Aunty Doll.

Acknowledgements

I would like to thank all the farmers for allowing me on to their farms and thus helping me to compile this book. I would like to thank them for giving me their permission to take and include photographs.

I would also like to thank our veterinary surgeons: Steve and Thaddeus at Midshire Farm & Equine, and arable specialist John at Frontier Agriculture.

My wife, Sarah, has been wonderful reading through the manuscript and taking on the enormous task of typing it up on to the computer.

Disclaimer

About the Author

John Terry was brought up by his parents on a private housing estate in Nuneaton, Warwickshire. At weekends and during school holidays he stayed with his uncle and aunt on a 500-acre farm in Leicestershire. His uncle was a farm manager and John, from a very early age, enjoyed all aspects of farm and country life. John left school at 18 and went to work on another farm before attending a 3-year college course at Worcester to become a Rural Studies teacher. He went back to the school he attended as a pupil to teach Rural Studies and remained there for the whole of his teaching career, which was 25 years. He was head of department and taught agriculture, horticulture and environmental studies, setting up a thriving school farm. At the same time, he wrote four hilarious books about life on the school farm. In 1989 he purchased a field that had not been farmed for a number of years, with no electricity, no farm buildings and a very poor water supply.The field was soon cultivated, grassed and a poultry unit installed for free-range laying hens. Over the years, the poultry enterprise increased. In addition, both pedigree Kerry Hill and Derbyshire Gritstone sheep have been bred and shown, winning numerous championships.

John's enthusiasm, determination, intelligence, humour and excellent communication skills have been apparent in all his endeavours, as he has proven himself to be a good farmer, establishing a farm of which he can be proud. John is good at getting what he wants – he gained planning permission for a mobile home on his site, which he lived in before he got planning permission for a permanent bungalow. Over the years, his farm land has increased to 13.92 ha (34.5 acres), and now includes a field to grow arable crops. He has always had an integrated life, formerly close to his parents and now close to his wife, Sarah, and his children, Jonathan and Roseanna. His family are involved in everything he does.

Contents

Introduction

Field for sale.

This is the story of how I developed a farm from a 3.164 hectare (7.82 acre) field that had not been farmed for a number of years, turning it into a thriving business and lovely home for myself and then my family over 26 years. Whilst telling the story, I have also explained how to set up and look after other agricultural enterprises. This is a modern agricultural textbook that is intended to be informative but at the same time highly amusing – which is what you would expect from me – especially if you have read my other books. Hopefully it will set you on the road to becoming a successful first-generation farmer like me.

A 3.164 hectare field to be sold by auction on 26 June 1989 situated on the Leicestershire–Warwickshire border.

I was lucky enough to purchase the field for £21,000 which was originally part of Vale Farm. It had not been farmed for a number of years and was covered in weeds. There were no buildings on the land and no electricity supply. A water supply was connected to a neighbouring farm. It reached the field in old lead pipes with very little pressure and was really unfit for human consumption.

Looking through the gate to the first field.

The land is fertile in nature. It is a good, level field, regular in shape with well-defined boundaries and benefits from two road frontages. I bought the field with savings and therefore didn't need a bank loan or mortgage.

Neither my parents nor my grandparents were farmers, so I was not born into a farming family. My uncle, Ben, was a farm manager on a country estate, so I spent many happy holidays and weekends helping him on the farm. The lady who owned the estate, Mrs Mary Caroline Inge, lived in a large mansion house. My uncle not only looked after the farm but also tended the gardens, grew orchids for her and chauffeured her Rolls Royce and Lanchester. Later, Mrs Inge died, and my uncle managed for Lord and Lady de Clifford, who inherited the farm and bought three neighbouring farms.

I spent my childhood with my parents, growing up on a private housing estate in Nuneaton. By the time I was 15, I had kept rabbits, guinea pigs, bantams, tortoises, mice, hamsters, newts, frogs and, best of all, my dog Lassie. I showed my rabbits and guinea pigs, which I enjoyed doing. Next came cage birds, which I exhibited at shows throughout the Midlands and with which I achieved two first prizes at the National Exhibition of Cage Birds held at Alexandra Palace in London. But I was not a farmer – I kept these animals as substitutes for farm animals – there was no room for cattle, sheep or pigs on a suburban housing estate.

Me aged seven, sitting between my parents with Uncle Ben and Aunty Doll. Even at this early age, I wanted to be a farmer.

After leaving school, I worked on a farm full time for a year. I then attended Worcester College of Education (now Worcester University) and became a Rural Studies teacher, teaching agriculture, horticulture and environmental studies. I developed and was head of the Rural Studies department and a thriving school farm, but I still had this burning ambition to be a farmer. I wanted to own my farm and not rent it. I couldn't afford to go out and buy a complete working farm, so I started off with the one field.

Chapter 1

How to Obtain Land

In reality, farming is sometimes not what it seems: for example, if you read children's farmyard books where Farmer Brown's wife merrily collects the eggs from her five hens, placing the eggs carefully in a lovely wicker basket. In fact, it is much harder work than this, often involving long hours and hard physical labour. When you start, you may be in full-time employment and have to work on your land before and after work, at weekends and during holidays. If you breed livestock, you will experience sleepless nights calving cows or lambing ewes, often on some of the coldest nights in winter when most townies will be asleep in their beds.

If it is your dream to own or rent a farm or a field, you must work hard and be totally committed. The farm may even make you a profit. However, there is the story of the farmer who won the lottery and, when asked what he was going to do with the money, replied that he was 'to keep farming until it has all gone.' On the other hand, you never see a farmer on a bicycle!

If you want to farm successfully, you certainly have to be a very determined person. Before you start, you should have some idea of what you are going to do with your land. Is it a full-time agricultural business or a hobby farm? You may be married, or you may have a partner, and you have got to respect their feelings about your ambition and dreams. They could love this rural lifestyle, or they could hate it. You will certainly be working long hours, and if your partner is not interested, they will see less of you than if you worked normal office hours.

Keeping sheep – I have two purebreeds here, Kerry Hill and Derbyshire Gritstone, plus some Texel crosses.

The main agricultural options or enterprises are as follows:

1. Livestock – the main enterprises are dairy, beef, sheep, goats, pigs and poultry, plus a few more unusual animals such as alpacas. A variety of these enterprises can be kept, or you can become a specialist in one or two areas such as poultry and sheep. You can also choose between keeping pedigree livestock and selling most of these as breeding stock or commercial livestock, which sell for meat production.

2. Arable farming – you need a large acreage for this if it is your only enterprise. You will obviously not make a living growing wheat, barley or potatoes on a small field, but hopefully you will make some money.

3. Mixed farming – this is a combination of livestock and arable.

Arable farming – combining our wheat.

4. A farm shop – selling produce on the farm, some or most of which will have been grown or reared on the farm. When you own a farm shop, you are tied down, and you are committed to opening hours. If you are married or have a partner, he or she may want to work in the shop while you are feeding livestock or working in the fields, but they may not. It is difficult for the two of you to go out for the day, and if you have young children, the problems extend to childminding while both parents are busy. If you close the shop, and no one is present on the premises, customers will turn to other sources if yours is not reliable. You can, of course, employ staff for your shop if you can afford it.

Farm shop.

If you are a first-generation farmer, you will probably start off in a small way, and then when finances allow, your enterprise will get bigger. Before you spend thousands of pounds on an enterprise, you must do your homework to make sure you can sell your produce.

The best and easiest way into farming is, of course, growing up on a farm that is owned by a parent, who would probably have inherited it from their parents. You would have been brought up on this farm and helped when you were a small child gradually learning the ways of farming until you could do jobs like milking cows or driving tractors and operating machinery independently. Many farmers' sons and daughters attend an agricultural course at college and sometimes work on other farms to gain experience including working abroad. Some sons or daughters would eventually become partners in the farm, or they would take over the farm when their parent retires. If there is more than one son or daughter interested in the farm, questions may arise as to who lives in the farmhouse or who is in charge of the farm. If the farm is to be shared between brothers and sisters, one or more of these may want to sell their share, which could become expensive for the remaining partners. The ideal situation is that you are the only son or daughter, and you want to farm the land. When I was a small boy, I remember watching a western film on the television. A cattle rancher took his son to the top of the hill and said in a real American drawl 'One day, son, this will all be yours – as far as the eye can see.' I can remember feeling really jealous! You could marry into farming, of course, preferably marrying an only son or daughter to avoid any inheritance complications!

During my time as a school teacher of agriculture and horticulture, I taught a few farmers' sons and daughters, many of whom worked hard and, after leaving school, worked in agriculture or jobs connected with agriculture. However, a student called Eleanor springs to mind. Eleanor had a real interest in the school farm, and she helped to show our sheep at agricultural shows. In her last year at school, she was not working hard enough during lesson time, and I challenged her and told her that she needed to improve if she wanted a career in agriculture when she left school. Her answer was quite simply: 'People with smart asses don't need examination passes!' She then stated quite positively, 'It's okay, sir, I'll just marry a rich farmer!' About 10 years later, I was showing sheep in Derbyshire, and there was Eleanor, holding hands with a young farmer. 'This is Raymond,' she proclaimed. 'We got married two years ago. Raymond is in partnership with his father, and they own 800 acres. We have a large dairy herd, beef, sheep and arable, and we are just having a new farmhouse built.' This is a true story and proves that it can be done but hopefully not at the expense of a loveless marriage. Marry the person not the farm!

Finding a partner who has the same interests and ambitions as you do is not an easy business. You are unlikely to find this partner – either in marriage or in business – in a pub or club or in the queue at the supermarket. Most of the population in Britain are townies and know nothing about the British countryside. If you ask most people today, they would not know a Holstein from a Jersey, or even an oak tree from an ash tree. The majority of the population are now disconnected from the countryside, and I am constantly appalled at the lack of knowledge that exists in the general public. After attending my rural studies lessons, my students had a good understanding of the countryside that would last a lifetime. Sadly, when I retired from teaching, my school farm finished and is now a wasteland. I can't help thinking that all the richness of education that the countryside can offer is being neglected in favour of other subjects that are transitory in our lives. I have employed older teenagers on this farm, and they know nothing at all about farming or indeed the environment outside their own front doors. Their parents bring them in cars, and they are not encouraged to look further afield than their iPods or PlayStations. I once told the headmaster at my school that a parent had kindly offered to knit me a jumper from one of my Kerry Hill fleeces. His reply was 'What a shame you have to kill a prize-winning sheep just to get a jumper!' It just goes to prove that if you are brought up with nothing to do with the countryside, you will probably not be able to engender an interest in your offspring. I realise that we can't all know everything about all subjects. I have a friend who can tell me instantly about any aircraft that happens to fly overhead, whereas I know nothing at all about aeroplanes. I was lucky enough to have relatives who were deeply involved with, and very knowledgeable about, the countryside, and I was blessed with parents who were willing to encourage and support my interests.

If you are 26 years old or under, the Young Farmers is a good meeting place. Sadly, membership has fallen, and some clubs have closed, but some are still thriving, and you don't have to be a farmer or have a job connected with farming to join. I was a very active member of the Nuneaton Young Farmers' Club and, over the years, became treasurer, chairman, club leader and a member of the advisory committee when I became too old to be a club member, and finally I was the club president. It is the youth movement for the countryside with the motto 'Good farmers, good countrymen, good citizens.'

Country Link is an organisation you can join after you are too old to attend Young Farmers meetings. You need to be 26 or over and be interested in the countryside and rural activities. They have regular meetings and visit places of local interest. It is stated in their advertisements that couples and single people are welcome, but Country Link is not a dating agency; nevertheless, it is a good place to meet like-minded people.

You can advertise for a farming partner. After leaving school, I worked on a farm full time for a year. I asked my farm boss how he obtained his farm. 'Oh I advertised,' he said with a straight face: '"Farmer seeks a farming young lady with a view to possible marriage. Please send a photograph of the farm and tractors".' He roared with laughter, adding, 'When I was 20, I couldn't keep my hands off the wife – so I sacked the hands and bought a tractor!' A friend of mine, Harry, had a text from the dating agency that said: 'Your advertisement has been on our website for the last eight years and you have still had no replies – do you want us to try it for one week without a photograph?' On a serious note, you can place an advertisement preferably in a country or farming magazine advertising for a partner with a view to marriage, or you can advertise for a business partner with capital, or a combination of these two. Alternatively, you can answer advertisements that have been placed in these magazines, or of course, you can find most things on the internet.

Renting

You can rent a farm complete with house and buildings. Many are privately owned and are often part of a large country estate; for example, the Crown is a massive landowner and takes on many tenants.

There are not as many county council holdings for rent as in previous years because much of the land has been sold off. Some counties have kept their tenancies on (60 county councils have council farms in England and Wales) and encourage farmers to get their first step on the farming ladder – Staffordshire being one of the best. They have 102 council farms in the county. They have starter units for new people entering the industry. The applicants must prepare a business plan and show what they will do with the farm. The average age of a new farmer on their county council farms is 32 years. The lease on a starter farm will probably be 10 years, and then after 10 years, if they have made a good job and want to continue, they may be offered a larger farm to rent, possibly doubling their acreage and being offered a lease for about 16 years. County council farms often have good buildings for dairy cows, and this is one of the best ways to get into dairying.

When a holding comes up for rent, it is advertised in farming magazines and on the internet. There could be over 100 applicants, possibly more. These tenancies are called Farm Business Tenancies, and they vary in the amount of time you can stay from 5, 10, 15 or 16 years or sometimes longer. Life tenancies are now very rare. Landlords will be looking for practical experience from you which could include a college course. You will also need capital, and they will try to get the highest rent they can. Rent will vary – a hill farm in Wales will cost less per hectare than an arable farm in Cambridgeshire.

Renting a field is much easier to come by than renting a farm complete with house and buildings. This is how many people get their first foot on the farming ladder, starting with one rented field and then finding some more, and then hopefully a field with some buildings. The field could be yours to use for just spring and summer, or autumn and winter, or long term. Many of these rented fields, however, are rented to people with horses and ponies because they are willing to pay a rent that is much higher than an agricultural rent – so you may find it difficult to compete with these people. Dairy farmers will often let out grass keep (which is easier to get than renting a field full time) in the late autumn when the cows are indoors, and they want grass grazing off by sheep. You need to be careful, because if you haven't got a proper agreement, the owner could insist you take the sheep off the land at any time, and you could find yourself with nowhere to put your livestock. You may pay a fixed rent for the land or so much per head per week. When you rent, you may enter into a contract where you have a grazing licence, which enables you to use the field for an 11-month or 6-month period – a break may then be taken between 11 April and 6 May – landlords traditionally like tenants to have a break. The land may have water connected or not. If not, you will have to carry it in. If it is a dairy farm, it is unlikely to be fenced well enough for sheep, and so you would have to invest in some electric fencing and a battery fencing unit.

Buying

You can't beat buying your land, but obviously you need some capital, which you may have had left you, or you have worked hard and saved your money. To go it alone and start an agricultural business, whether it be part time or full time, you really need to know about book keeping, selling and marketing, and talking to people. One in three new businesses in Britain fails in the first 3 years, often owing to not marketing the goods or services well enough, cash-flow problems or encountering customers that are bad payers (I know of a very good business that failed because the owner just could not get in the money he was owed). Partners can fall out – perhaps one partner wants to leave, or one partner does not work hard enough or takes too much money out of the business.

Then, of course, there is the great British weather, falling prices, rising costs of animal feed, pests and diseases, the bank calling in the loan and running out of money completely. The family can also be upset because you do not have enough time for holidays and weekends off. I was lucky my father was an office manager, and he did the accounts for me and then taught me how to do them. I have always been good at selling and could sell a lawnmower to someone without a lawn, and being a school teacher I am good at talking to people and dealing with farm staff. When dealing with staff, don't leave everything to them – keep your finger on the pulse; watch and observe! Once you have decided what enterprises you are going to carry out, even if it is part time, you need a business plan, which you may need to show to the bank if you want a loan.

Buying second-hand buildings and equipment will save you money. If you run short of money, you might want to borrow from members of your family but any borrowings must be put in writing to save any falling out. If you decide to sell at the farm gate, you will need a website, which should look professional to sell your produce well.

Once you have rented or bought your field, you will need to consider the following things.

If you are completely on your own, you are known as a sole trader. You need to register your farm business with Her Majesty's Revenue and Customs (HMRC) and file a self-assessment tax return each year. An accountant will fill the forms in for you. Sole traders and business partners are self-employed, and if they make enough profit they must pay income tax. If you are self-employed, you normally have to pay class 2 National Insurance Contributions. If your profits are over a certain amount, you also pay class 4 contributions – check the HMRC website for details.

A limited company has directors and shareholders, and the advantage of a limited company is that it is the company that is responsible for debt, and not the owner. You will need to register for VAT if your annual sales go above a limit of £79,000 (figure correct as of 2014). VAT is a tax charged on most business transactions. Businesses add VAT to goods and services. The standard rate for VAT in 2014 is 20%. It is possible for the farmer to claim VAT back on many items. 'Free' money is rare but you could always try. You can possibly get a loan that is a government-backed Start Up loan – check the 'Start your own business' information on the government website for up-to-date information.

A New Enterprise Allowance is currently available for people aged 18 or over with a business idea that could work. The allowance can provide you with money and support to help you start your own business if you get certain benefits. You must check the website to see if you could qualify for help. You could get a weekly allowance paid for up to 26 weeks, up to a total of £1,274 (2014), and you could apply for a loan to help with start-up costs. The loan must be paid back, although the allowance need not be. A business mentor could also be available to help you.

The Prince's Countryside Fund gives grants to support the people who care for the countryside. It supports rural enterprises and farming businesses, providing training opportunities for young people and educating people about the value of the countryside. The majority of the funding comes from supporting companies.

The Basic Payment Scheme (BPS)

The Single Farm Payment was an agricultural subsidy that was paid to farmers in the EU. It has now been replaced by the Basic Payment Scheme. The Rural Payments Agency makes payments, traces livestock and carries out inspections. To receive the Basic Payment Scheme, farmers have to comply with a set of Standard Management Requirements called Cross Compliance. These rules keep the land in good agricultural and environmental condition. If you disobey any of these rules, you could lose a proportion of your Basic Payment. Standards would include management of soil erosion, and protecting and managing water. To get the full set of up-to-date rules and regulations, check the Rural Payments Agency website. This Basic Payment will be taken into consideration in any agreed tenancies – sometimes the landlord receives it, and sometimes the tenant. It represents a large proportion of income for many farmers and subsidises farmers on a per-hectare basis. Many farmers claim they could not make a profit without it. Beware: you could have a good tenancy, and you could have worked hard to improve your rented farm. You may have built up a dairy herd, for example; then when your tenancy is up, your landlord decides not to renew it, and you then have a dairy herd and nowhere to go!

If you have claimed Single Farm Payment you will have existing entitlements to payments. These entitlements are now your BPS entitlements. The entitlements cannot be used unless you are an 'active farmer' and have at least five hectares of eligible land and at least five entitlements. Common land is also eligible for BPS. If you are a young farmer or a new farmer and have no entitlements, you must provide evidence of your agricultural activities to support an application for entitlements from the national reserve. Check the CAP website for up-to-date information.

Greening Rules

The BPS will include new 'greening' rules, which you must follow in order to receive your full payment – this is worth approximately 30% of the money paid to you. The term 'greening' refers to helping the environment and the climate. Organic land automatically qualifies for the greening payment.

Farmers with over 30 hectares of eligible arable land will usually need to grow at least three crops (the largest crop must not cover more than 75% of the arable land). Farmers with smaller areas of arable land (10–30 hectares) will need to grow at least two crops (the largest crop must not cover more than 75% of the arable land). Farmers with more than 15 hectares of eligible arable land must now set up an Ecological Focus Area (EFA) which must be equivalent to 5% of their total eligible arable land. There are, however, some exemptions – check the Rural Payments Agency website.

So, farmers now have to sit down each year and work out how the farm is going to meet the new greening rules. Areas to consider and count as part of your EFA are:

1. Buffer strips. These must be next to a watercourse or parallel with and on a slope leading to a watercourse ('watercourses' include ditches). These strips of land must have a minimum width of 1 metre, and must have no food crops grown on them – but they can be grazed or cut or sown with wild bird seed. Buffer strips must be distinguishable from neighbouring EFA fallow land.
2. Areas in which you grow nitrogen-fixing crops. This includes beans, peas and lupins. The minimum area which can count as part of your EFA is 0.01 hectares.
3. Well-managed hedgerows.
4. Areas in which you grow catch crops or cover crops. The minimum area which can count as part of your EFA is 0.01 hectares. Crops could include barley, lucerne, mustard, oats, phacelia, rye and vetch. You need to sow two different cover crops mixed up together – one cereal and one non-cereal. They need to achieve ground cover, utilise available nutrients, and establish quickly.
5. Fallow land. However, wild bird seed and plants with nectar to encourage pollinators can be sown on fallow land.

I have just summarized the rules – you must go into it in much more detail, and you will need to check the Rural Payments Agency website for all the up-to-date rules and regulations.

Chapter 2

Buying Your First Field

First, and above all, you need some capital. I was fortunate in that I didn't need a bank loan or mortgage to buy my fields, but I wasn't left the money in a will – I worked hard to get it, working on farms, teaching, and giving talks and after dinner speeches. The Agricultural Mortgage Corporation are the specialists in providing mortgages for established farms, equestrian rural business and horticulture. However, they are probably not the company for you because, for example, if you lived in a three-bedroomed house with a mortgage, on a normal housing estate, and you wanted to buy your first field, they would not be interested in lending you money. They will only lend to you if you have an existing farm showing 3 years of healthy accounts and showing a profit, which is not encouraging for the first-generation farmer, so to borrow money in this situation, you would have to go to the bank or the building society. Generally, the smaller the parcel of land, the more money it fetches per acre or hectare. A 2 acre (0.809 hectare) paddock suitable for ponies will cost a lot more per acre than a 100 acre (40.46 hectare) arable field. It may be your ambition to build a house on it one day and so live on the premises. If it is a 1 acre field, then this is unlikely – you stand more chance of achieving this with a larger field or fields. I really think this is an excellent option to get on the farming ladder – as long as you have enough money or you can get enough money to buy.

When you become a land owner, you have rights; if you own the land outright, you are not going to be forced off it by a landlord – admittedly, if you owe the bank and can't repay your loan, they could call in the loan, but if you keep up your payments, you will be safe.

Agricultural land is sold by rural estate agents. If you want a field or fields in a certain area, you must contact these estate agents to see if any land is for sale. If not, ask to be on their mailing lists and email lists, and then if a suitable field comes up for sale, they will send you the details.

Estate agents will put a 'For Sale' sign up on the edge of the field, in the gateway or in the hedge. This is how I found my first field. Land is advertised every week in farming magazines such as 'Farmers Weekly' and 'Farmers Guardian' and on the internet. You will probably have to be patient to find something suitable and in the area you want to be in. The following is what you need to look out for:

Guide Price – e.g. £55,000–65,000 of Pasture Land Extending to 6.4 Acres

1. Tenure – E.g. freehold.
2. Situation – The property may have planning permission for buildings or a mobile home, or just outline planning permission. This will, of course, make the property more valuable. If the field adjoins the edge of a village or town, the chances of building a house are enhanced, but the chances of putting up a pig unit may be reduced, as this would be very unpopular with the local residents. Fields that are situated in areas of outstanding natural beauty or land in National Parks or conservation areas will cause problems when you try to get planning permission for buildings and a house. You will be restricted on the materials that you can use, the siting of the buildings and the architectural style.
3. Description of the property – A good loam soil with plenty of depth is ideal. Rocky ground with only a thin covering of soil will not grow a good crop and would also be difficult to cultivate. It would be expensive to dig out for buildings. A steep hill is again difficult to grow crops on and is dangerous for use by tractors and implements. A hill farm would cost less per hectare than a lowland farm but would probably not have such a good depth of soil or kind climate, and so you will be restricted in what you can grow and what breeds of livestock and the number of livestock you can keep. If the land is prone to flooding, then it is probably not advisable to put buildings or a house on it. The Environment Agency website will show if the land is in a risk area. These days, with very wet winters, check the district to see if it floods. It must be heartbreaking to sow a crop and see it ruined.

Flooded farm land February 2014

4. Access – Land is best accessed from a road; road frontage is a definite bonus, but beware of dangerous bends in the road. These could affect planning permission for a house at a later date. Some land can only be accessed by a right of way. The right to use this must be formal and not just a verbal agreement. I have been lucky that all the fields I have purchased have had road frontage.

5. Services – Many fields are sold with no services on the land. Land connected to water and electricity will be worth more to you than land not connected to existing utility services. It is unlikely that an isolated field will be connected to sewage pipes.

6. Plans and particulars – The catalogue that is sent out to you from the estate agents will include a plan of the land, and they usually say 'It is believed to be correct in every way,' which covers their backs in case there is a mistake. Boundary ownership, where known, is indicated on the maps by an inward 'T' shape.

7. Wayleaves, easements and rights of way – An easement is a right over one piece of land existing for the benefit of another piece of land, e.g. British Rail may have an easement over part of the land for maintenance and repair of the railway, or a neighbour may have the right to use a farm drive. If the electricity company has poles that stay erected in the field, the owner of the field will probably receive a wayleave payment for allowing the poles and stays on the land. Take note of public footpaths and bridleways that cross the land – I have none.

8. Uplift clause – Sometimes the land is sold subject to an uplift clause, which might be 25% should planning consent be granted for either commercial or residential use over a period of time – perhaps 25 years from the date of sale. This means that the vendor has a legal right to a share in the increase in value of the land. In my opinion, this is not fair – once you sell a property, you should sell it lock, stock and barrel, and that's it. Many would not agree with me – certainly not the sellers – but buyers don't like it, and it puts many off purchasing the land.

9. Restrictive covenant – Beware: this could include 'Not at any time hereafter the sale to build, erect or place or allow to be built, erected or placed on the property or any part thereof any hut, tent, temporary dwelling, caravan, house on wheels or encampment intended for use as a dwelling.' This is not good if you want to build up a farm.

10. The new Basic Payment Scheme – Entitlements may or may not be included in the sale. If you are buying privately, don't forget to ask for them – land is worth more with these entitlements. If the land is sold by auction, details of these entitlements will be included in the catalogue.

Local Authority

e.g. North Warwickshire Borough Council plus a telephone number.

Vendor's Solicitor

The vendor's solicitors prepare legal information packs including searches, title deeds and Land Registry entries. If you need to borrow money, your bank or building society will ask a surveyor to prepare a valuation report. Your solicitor will need to carry out all your legal requirements, and you will have stamp duty and legal fees to pay. Don't forget the old farming question: 'What's the difference between your bull and your solicitor? The solicitor charges more!'

Viewing

If it is an isolated field, viewing is often during reasonable daylight hours with a copy of the particulars to hand – this being an authority to view. It is essential that you visit the site before purchasing, as it may be a disappointment when you see it, or it might be just what you are looking for. If possible, take an experienced farmer with you to get a second opinion. Look at the guide price and then work out what you think the land is worth to you. Work out how much you can afford to pay and in your mind decide what your top price will be. You will probably need to go to the bank or building society to arrange a loan, and you will also need to sort out your savings – putting them all into the same account – you will not be allowed to pay cash. If the land is overgrown with weeds, it might put some buyers off. Old and derelict buildings may also discourage some, but it may be a good thing, as it may be easier to get planning permission for new buildings where the old buildings stood. The field or fields may have a history of failed planning applications – you can check this out by visiting council offices' planning departments. The owners of the first field that I purchased had applied to build a house on it, and the application was refused, which was not surprising, as there was no agricultural business on site, and it is a greenfield site.

The National Planning Policy Framework does change from time to time, so check out their website for up-to-date information. Also, you can check the Land Registry – they record land sold, and you can find out what has been sold in your area. If you go and locate this land, you can see what has been done with it. If land is sold privately, the asking price may be included in the advertisement, or you may have to contact the estate agent to find out. The price could be negotiable, so you may be able to put an offer in just as you would when buying a house. Note that the guide price is not the figure that the property will sell at – just a guide.

Informal Tender

Land can be sold by informal tender. Bids are submitted in an open or sealed envelope. Envelopes are opened by the estate agents as they receive them. After making your bid, the estate agent may contact you after the closing date and ask if you would like to increase your bid because they have had higher bids than yours.

Formal Tender

Formal tender is another method of buying land. Offers are returned in a sealed envelope and are all opened together on a published date and time. The highest bidder buys the land and will have to complete the sale on the contract-appointed date.

An auction in progress.

Buying at Auction

Buying at auction is an excellent way to obtain land. Most lots offered for sale are subject to a reserve price. This price is agreed with the vendor and auctioneer, and is confidential. You may also have asked a builder or architect to survey the property. I have bought twice at auction, and believe me, your pulse rate will certainly rise! You will have completed your site visits, decided what you can do with it and sorted out your financial arrangements before sale day. You may want to make an offer before sale day, which may be accepted, in which case the sale would be cancelled. This offer must be in writing or by email or fax. Many vendors would not accept offers before the auction. On auction day, arrive in plenty of time. If you are the successful buyer, you will need to pay a deposit of 10–20% of the selling price on the day and pay the balance usually within 20–30 working days. You could lose your deposit if you fail to complete the sale on time.

The land that you are interested in probably won't be the only lot for sale. There will be a number of properties auctioned that day. I had to wait until the end of the sale to buy my two fields at auction sitting through the sales of complete farms and country houses – the waiting is very agonising. I like to sit at the back of the room in the centre. From this position, you can see what is going on, and at the same time you are out of the limelight. At the start of your lot, the auctioneer will announce if there are any changes to the details shown or anything to add. He may talk about similar lots that have recently been sold. He will praise them up if they have reached a high price, encouraging you to bid. The use of favourite phrases such as 'This is a once-in-a-lifetime chance to get this land' really help to get your pulse racing. The vendor may bid, or the auctioneer may bid on behalf of the vendor up to the reserve price but may not take a bid equal to or more than the reserve price. I don't open the bidding even when I am buying sheep at auction. Sometimes the first bids will be false which may be made by the auctioneer's staff – this helps to get the auction underway. You must concentrate and make sure the auctioneer has got your bid. You may see others bidding, but some will bid discreetly with perhaps a nod or a wink. Keep your concentration and your cool, and make sure you don't bid against yourself.

If you are lucky enough to have the last bid, the auctioneer will probably say 'I'm selling once' and pause, 'I'm selling twice' and pause, 'I'm selling for the third and last time' and then shout 'Sold!,' and the hammer will go down. The land is yours, and you have entered into a contract. In the excitement of the bidding, you may want to keep going above your maximum price, but it is important not to get carried away and then find you can't pay. On the other hand, don't lose the property over a small amount. If the reserve price has not been made, the lot will not be sold. If you still think you can afford the property, you need to see the auctioneer straight after the sale – don't leave it even 5 minutes, because you may lose it to someone else.

When you have purchased your property, you will feel excited and relieved, and almost have to pinch yourself to believe what you have done. You will also wonder whether you have done the right thing and worry about how much money you have just spent, but on the whole you will feel elated. Your property will need insurance as soon as you have exchanged contracts, as it is now your responsibility, especially if it has a house or buildings on it. Land is a good investment. I paid £21,000 for 3.164 hectares (7.82 acres) in June 1989 – £6637.16 per hectare, £2685.42 an acre. In February 2009, I paid £77,500 for 3.273 hectares (8.09 acres) – £23,678.58 per hectare, £9579.72 an acre.

When you have bought your land, you need to check that it has a holding number. If it is a complete, established farm, it will have one. If it is an isolated field, it will probably once have been part of a farm, and you can't use the old holding number, as the rest of the farm will still be using it – you will need a new number, which you can get by registering over the telephone or filling in a customer registration form at the Rural Payments Agency. They will need to know the full postal address of the field, but if it hasn't got one, you need to give them a grid reference. The holding number is made up of three lots of numbers – the first lot is the County, the second lot is the Parish, and the third is your unit holding number, e.g. 21 211 0041.

Chapter 3

Services: Electricity and Water

Mains Electricity

When I bought my first field in June 1989, there was no electricity on site. I was eager to start my free-range poultry enterprise. I bought the building, and contractors erected it and fitted it out with all the equipment. I purchased the pullets, and I was up and running all without mains electricity.

I purchased a second-hand generator, which was powerful enough to run the lights, fans and chain feeders, but there was one problem: the fuel tank would not hold enough fuel to keep the generator working throughout the night. I was living with my parents at the time, so every evening at 10 pm I travelled the 2½ miles to the poultry unit, filled the generator and went home to bed. My father was good enough to get up early in the morning and drive up to the poultry unit to fill the fuel tank before 6 am. I had applied for electricity to come on site before the poultry unit was even built, but I had to wait about 3 months for it to be supplied. Today, my local electricity company tells me that the timescale depends on your location and how far away you are from the existing supply. You will need to be connected to the nearest suitable transformer or substation. You have the option of the supply being brought to your land on poles or the more expensive option of underground cables. If you choose to go for the underground cables, the electricity company will do all the work, but you are allowed to dig out the trench yourself or

Electrician connecting my electricity to the poultry unit.

with contractors, which will probably be less expensive. Before you start digging, the electricity company will come out on a site visit and show you the best place to dig, and of course, you need to avoid existing underground cables and pipes.

To get mains electricity onto my land, I chose the less expensive poles option. The company carried out all the work, which cost £5000 in 1990.

We then had to build a small shed to house our electrical fuse boxes and meters and dig a trench from the last pole to the box to lay cable. We also needed a trench to lay cable from the box to the poultry unit, so all of this, including the equipment, cost another £1000. This last lot of work was carried out by a local electrician, who was excellent. We have employed a few electricians since then (including the present one, Stuart), who drive Volkswagen vans with the 'Volkswagen' crossed out on the back and replaced with the word 'voltswagen.' Stuart also tells me he is up to date because he is a current specialist. Each year, I receive a wayleave payment for each pole and stay that stands on my land – seven poles and two stays only bring in a cheque for £72.44 (2014).

Water Supply

Water flowed very slowly down old lead pipes from my neighbour's farm to my first field. It was metered, so I had to pay my neighbour twice a year, and then she paid Severn Trent Water. I wouldn't have dared drink the water, as it often had bits of metal in it. I ran my poultry unit using this water for a while, and I never had any trouble with the hens dying after drinking it, but it is advisable to flush the drinking-water through with an approved chemical occasionally to stop the build-up of bacteria, which could affect the health of your hens. When I had enough money, I was connected to the mains water supply in anticipation of building a dwelling. If you are lucky, you will buy a field or field plus buildings or even land including a house, which will have a mains water supply. Make sure you know where the stop tap is – it is usually under the sink in a dwelling. Your supply pipes could be old lead pipes, and if so I would advise getting them changed to plastic. Lead pipes are not used these days because they are considered to be harmful to health. If you decide to replace any lead pipes on your property, it is your responsibility to fund the work. The water company will connect the new pipes to their network and replace their lead pipes if necessary. Lead pipes are a dull grey in colour and are easily identified; however, renovating lead pipes in your house could leave your home unsafe. Years ago, lead pipes were used as an electrical earth, so ask your electrician for advice before going ahead.

Your water pressure could be low for a number of reasons:

1. if the water supply pipe is not just for you but is shared by you and neighbours;
2. the height of your property above the water main;
3. peak demand – when many people are all using water at the same time;
4. the layout of the water pipes inside and outside the buildings;
5. possible leaks.

If you have a water meter, it is easy to check for leaks. Turn off all taps, and then wait for half an hour. Read your meter. Keep your taps turned off for an hour, and then read your meter again. If the reading has increased, you have got a leak – you will also see the numbers moving on the meter. If the leak is on your property, it is up to you to repair it. Check to see if your pipes are lagged, especially in unheated barns, buildings and outside at the water tanks and external taps. Lagging is important, as frozen pipes mean that livestock are not getting their water. Burst pipes make a mess, and it is extra work repairing them.

Contractor digging out a trench to lay water pipes for my poultry unit.

Bringing Water Onto Your Land

To bring water onto your property, you need to apply to the water company. Severn Trent Water is the company that provides water in my area. They will inspect the site and provide a quote for all the work that needs to be done. With Severn Trent Water, you have the option of digging out the trench and laying your water pipes from the nearest water main to your taps. They will recommend 'water marksmen' – contractors who are trained by Severn Trent and are qualified to do the job, or you can use your own contractors. It is probably less expensive to have contractors to do the work rather than the water company. The pipe needs to be buried at least 1 metre deep so that it is safe from frosts and implements such as ploughs.

You need to buy the water piping and then pay for the labour. If you do this, the water pipes then belong to you, and if you get a burst pipe or another problem, it is your responsibility to fix it. After you have installed your water pipes, Severn Trent Water will purchase them from you, and then they are responsible for them. They will, of course, only purchase piping that has been laid outside your property, e.g. from the village to your boundary. Once the water pipe reaches your boundary and gets onto your property, it becomes your responsibility. You will need your water company to link up your new pipes to the nearest water main. At the end of 1992 and the beginning of 1993, I employed contractors to bring a water supply from the village, which is half a mile away, to my poultry unit. The plastic pipe was laid in a trench dug out by a JCB. The biggest problem was that the grass verge was very narrow in places, and pipes had to be brought under the road to get to my property.

Severn Trent Water connecting my new water pipes to the mains water supply.

I kept my water pipes – I wasn't given the option to sell them to Severn Trent Water. The water meter is in the village next to the water main, and so piping from the water meter to the taps is all my responsibility. I keep a close watch on it, as I don't want others tapping off the supply. Back in 1992, all the work, including Severn Trent connecting me up, cost a total of £6000.

Chapter 4

Hedging and Fencing

Planting a Hedge

A young hawthorn hedge growing nicely.

I have never had to plant a hedge that ran the whole length of a field, but I have planted and filled in gaps where the hedge had died and I also planted a hedge around my garden when I lived in the mobile home. I enjoy planting hedges. They are an important part of the countryside, providing a wonderful environment for plant and animal life. A hedge also acts as a corridor so that animals and birds can move up and down with plenty of cover, but as a farmer they are valuable in other ways. Hedges make a good windbreak and shelter. Cattle and sheep will huddle against the hedge in bad weather and take shade in hot weather. Hedges stop livestock from straying, act as field boundaries and can be used to hide buildings and eye sores.

The most common hedging plant is the hawthorn. Hawthorn dominates on this farm, and mixed within the hedgerows are blackthorn, field maple and hazel. The trees growing in the hedges are nearly all oak and ash. I had no hesitation in filling in my gaps with hawthorn and some field maple. To plant your hedge, you can spray with a weed-killer first to get rid of all the weeds. 'Roundup' is a good weed-killer for this. If the ground is hard, it will need cultivating, but you may only need to do a narrow strip. To fill in my gaps, I hand-weeded and then dug the strip of ground over with a spade. It is best to plant your hedge when the plants are dormant – from mid autumn to late winter – making sure the ground is not waterlogged or frozen. The plants that you buy from the nursery will be bare-rooted; they are dug up from a field and are not container grown – container grown plants are more expensive. When you fetch or receive your plants from the nursery, you will probably find they are tied up in bundles. They are not expensive – hawthorn (*Crataegus monogyna*), 40–60 cm high and 2 years old, would cost about 34 pence each. If you buy more, they would cost less. Field maple (*Acer campestre*), 40–60 cm high and 2 years old, would cost about 35 pence each. Again, they will cost you less if you purchase more (2014 prices). You must not let the roots dry out, so if you have hundreds to plant, heel them in, and use them when you are ready. Heeling in just means digging a trench and planting them in the trench close together.

When planting, keep the plants in a bag, and take each one out when you are ready to plant it. This will prevent the roots from drying out. Use garden lines to keep your rows straight, and a pair of gloves is advisable to protect your hands from thorns.

Plant a staggered double row 30 cm between the plants and 30 cm between the rows. Don't dig out a hole for each hedging plant; just push your spade into the soil, pull it backwards and forwards to make a slit, and then plant your hedging plant in the slit. Remove your spade and press around the base of the plant with your foot making good root contact with the soil. Each plant should then be protected from rabbits with an individual spiral guard. A cane can also be placed inside the guard to give extra support. Water them if you can. When you have finished, look back at what you have done – you should be very pleased with your efforts. You can mulch the hedge with wood chips to keep the moisture in, but this is not essential. Stop weeds from smothering the plants, and don't prune them for a year. After a year, you can cut the tops off. If you do this every two years, cutting higher up the plant each year, they will become bushy plants.

Cutting and Laying a Hedge

Hedge cut and laid.

This is a skilled job, and it is no good reading a book and then having a go at it. You need to go and learn the skill from a craftsman before you cut and lay your own hedge. You may be able to attend a specialist college course, or you can get the craftsman to come and cut and lay your hedge for you. My uncle Ben, the farm manager on a large country estate, was a master at the job and spent a great deal of the winter months cutting and laying hedges. He taught me how to do it, but I am not a craftsman at hedge laying. I have done a little in the past, including a 48-hour hedge-cutting marathon as part of a charity event organised by Nuneaton Young Farmers' Club.

Since buying my fields, there has only been one hedge that needed cutting and laying – a hedge growing in a bank at the side of a road. It was too much for me to tackle, and so I employed contractors to do it, Matthew and David.

If hedges are just left for years on end, they look like trees all growing in a line. They put all their energy into height and lose the thick, bushy growth lower down, making it easy for livestock to push through. A hedge needs to be thick and bushy, and cutting it helps to regenerate essential lower growth. The sides of hedgerows also need to be cut to prevent them widening and growing into the field reducing the available space for livestock and crops. To cut a hedge for laying, you have to cut nearly all the way through the base of the stems, and then these stems are laid over at an angle of approximately 35 degrees. After dropping the stems – now called pleachers – they are arranged parallel to each other along the length of the hedgerow. Old dead wood is removed, and to strengthen the hedge vertical stakes are knocked into the hedge about 0.46 metres apart. The stakes are about 1.7 metres high and have a diameter of about 50 mm. They are usually hazel or ash. Often, hazel is woven round the tops of the stakes. These binders are about 3 metres long by 40 mm wide, and they make the hedge strong. A well laid hedge could last for 50 years if it is kept trimmed before it needs laying again.

There are many styles of hedge cutting. The Midland style is the most common; one side is left thick with brush (bushy), and the other side is clear of brush. The livestock would be kept in the field behind the brush with the road or an arable crop on the other side. A top man laying a perfect hedge could lay 20–25 metres a day, but a good aim would be 10–15 metres a day. Nowadays, to remove a hedge, you need permission from the planning department of your local council authority. Previously, government grants were paid for farmers to remove hedges to make larger fields for bigger tractors and machines.

Mechanised Hedge Cutting

Nigel cutting my hedge with a flail hedgecutter. This field had just been purchased in February 2009.

Most hedges are now cut with a flail attached to the rear of the tractor. Cutting and laying hedges takes a long time and so is expensive. Flail cutting is quick and less expensive. The flails cut and break the branches and twigs into very small pieces, and so there is no need to clear up the cut timber. A contractor is best used to cut your hedge – it is not worth spending a large amount of money on a hedge cutter when you only have a few hedges to cut. Cutting the hedge encourages thick, dense cover. Most trees and shrubs flower and produce seed on 1-year-old twigs, and so if you cut your hedge each year, they will not have time to produce fruit or berries for wildlife. However, brambles and wild roses will produce fruit if you don't cut too early. Therefore, to be environmentally friendly, cut your hedge every 2 or even 3 years, and don't cut all of your hedges in the same year. It may be necessary to cut some hedges each year if they border a road or a footpath for safety reasons. Many farmers do have their hedges cut each year. Hedges are cut each year on this farm because I like the farm to look very tidy. I have plenty of neighbours who cut their hedges every few years.

If you are a Basic Payment Scheme (BPS) claimant in England, it is a rule or condition that you do not cut your hedges between 1 March and 31 August – with certain exceptions. Again, to be environmentally friendly, leave autumn fruits and berries on the hedge, and cut your hedge preferably during January and February. However, on this farm, they are cut earlier because by January, the ground would be too soft, and the tractor would make too much mess. Also, the hedge wants cutting in the arable field before the crop is sown; otherwise the tractor would do too much damage.

An excellent finish to my hedge cut with a flail.

The contractor, Nigel, who cuts the hedges on this farm is a highly skilled operator and has won many prizes for his hedge cutting.

Fencing for Sheep

Fences keeps livestock in and your neighbour's livestock out. They also act as your field boundary. Therefore, fences are important as this story shows. I once taught a knowledgeable farmer's son called David. His maths teacher asked him the following question: 'If there are fifty sheep in a field, and one gets out through the broken down

Scots pine fencing posts with sheep wire attached in the field I purchased in 1996 – the hedge has been cut on one side, but the top has not yet been cut.

fence, how many sheep are left in the field?' 'None' replied David. The teacher said 'You don't know maths do you?' 'Sorry Sir,' said David, 'but you don't know sheep!' A good practical fence can be easily erected with stock netting/sheep netting and Scots pine fencing posts. The posts are pressure-treated, and you can buy machine-rounded posts or peeled posts. I prefer the machine-rounded posts – they look neater and are very uniform. You can also use half posts. These are more difficult to drive in straight, and they will therefore lean at an angle. I fenced my first field in 1989 with half-round posts, and most of them are still in place today, but I fenced the other fields with machine-rounded posts. Before I started fencing, contractors cut the hedge back with a hedge cutter as far as possible – this gained me almost an extra metre of field all the way round.

Machine-rounded posts are 1.7 metres long with a top diameter of about 90 mm. About 0.7 metres will go into the ground – space them 3 metres apart. You need a stronger, thicker and longer post at the corners of the field called straining posts. You need to space these larger posts every 150 metres on a straight run followed by the smaller posts for another 150 metres and so on around the field. These larger posts would measure 2.15 metres long with a top diameter of 125 mm. This post would also need about 0.9–1 metre buried in the ground. At the base of these large posts, two long pieces of timber will need to be notched in at each side at an angle of about 45 degrees from the upright. These are known as godfathers, and they help to stabilise the upright post.

If you are dividing up a field, use a string or wire to get your posts in a straight line. If you are fencing around your boundary, just follow the line of your hedge if you have one. Lay all your posts out on the ground or lean them up against the hedge before starting to put them in. Use a measured length of timber (it could be a rail) to measure the gaps between the posts rather than having to keep unrolling a tape measure. Start knocking your posts in at a corner. You can use a sledge hammer, but this may damage the tops of the posts, or you can use a post rammer, or best of all use a tractor-operated machine. When all your posts are in, you are a ready to attach the wire. Use good-quality high-tensile fencing – this will be 80 cm high. The standard wire used on most farms today is C8/80/15. C8 means 8 horizontal line wires, 80 means it is 80 cm high, and 15 means 15 cm between each vertical wire.

Make sure the small squares in the wire are at the bottom of the fence. I have seen people use this wire in their gardens, and they have put it the wrong way up – I have also seen it used on a farm the wrong way up!

Start in a corner, wrap the wire around the post, and then staple it. By staple, I mean the double-ended nails shaped like a horseshoe and driven into the timber with a hammer. To pull the wire really tight between the posts, you can use a special straining tool or ratchet-type metal strainers, or best of all attach it to a tractor, land rover or other vehicle, and then pull. Barbed staples are better than plain ones, as they have less chance of coming out. Don't hammer the staples quite all the way in, as they could damage the galvanised wire, and it won't last as long. After all the wire is up, you then need to add a strand of barbed wire above it. If the fence is next to a thick hedge, one strand will be sufficient about 15 cm above your fence, but if the fence is dividing a field into two, two strands is better, or you can use one strand of electric wire.

Fencing for Cattle

Dairy farmers and beef farmers probably would not go to the trouble of using sheep/stock netting. They tend to use a simple fence of Scots pine posts with strands of barbed wire attached. The posts would again be placed about 1 metre apart. This type of fence will stop cattle getting into ditches and straying out of gaps in the hedge.

Post and Rail Fencing

Post and rail fences are certainly not necessary around the edge of a field next to a thick hedge, but they can be used for dividing a field into two, and they look really good on either side of your farm drive.

Post and rail fencing being erected on both sides of the farm drive in 1997 soon after the bungalow was completed.

Pressure-treated square sawn posts and rectangular rails, in my opinion, look the best. Another option is round posts or half-round posts with half-round rails; the posts and rails can be peeled and treated, and these vary in girth, or you can use machine-rounded posts and rails, which are all even. Yet another option is chestnut cleft: the rails fit into a mortice in the post, and it is rustic to look at and is hardwood – you can use hardwood or softwood posts. We fenced off our farm drive using square sawn posts and square rails. My posts are typical – they measure 125 mm by 75 mm and are 2.1 metres long – 1270 mm is left standing out of the ground. The rails are also typical: they measure 88 mm by 38 mm and are 3 metres long. You can use a smaller post between the longer posts (called a prick post), but I prefer all large posts, although this is more expensive. You need a post at the end of each rail and one in the centre of each rail. Use a string to follow to get a straight line and knock all the posts in first. A post driver behind a tractor is best for these large posts. Once your posts are in, you can staple on your stock netting if you need it and then finally attach your rails. The rails should be fixed on the field side where the livestock are, as it creates a strong barrier. However, I put my rails on the drive side not the field side. This is because it is more pleasing to the eye, and I only graze sheep in the field so they are not going to knock the fence down.

The second and bottom rail are cut in half and then nailed on. The top rail and third rail are left as full rails. Then, work your way along from one end by staggering the rails in this fashion. This will make the fence stronger. When you get to the end of the run, you can use the half rails that you have sawn to finish the second and bottom run of rails. Make sure you butt the rails up to one another and that they run parallel. Galvanised 100 mm nails would be sufficient to secure them to the posts.

Post and rail fencing completed.

Electric Fencing

Electric netting keeping my hens in.

Before I bought my first field, I kept my pedigree Kerry Hill sheep on grass keep, and many of these fields were not fenced. I used rolls of electric netting (polywire) with an energiser unit. I have used a 12-volt wet battery and dry disposable batteries. You can, of course, recharge your wet battery. You can get solar-powered units, and you can also get an energiser that enables you to take the power from the mains. This is the best option if your fence is to go near a plug socket because you haven't got a battery to go flat. The wire running from the barn or building (where the energiser is) will obviously need to go outside to the fence. This needs to be installed safely where there is no danger of tripping over it. Energisers give a high-voltage, low-current pulse rate about once a second.

With sheep, you can use three strands of wire, but this would not stop my Kerry Hills – electric netting is definitely best. These mesh fences are portable and can be moved from field to field, and all the horizontal wires are electrified except the bottom line, which prevents the leakage of current going to earth.

A good earth stake is needed – a wire from the energiser goes to the fence and another wire to the earth stake. This earth stake is definitely needed; without it, the livestock will only get a very mild shock or no shock at all. I tip a bucket of water on the earth stake, which then runs into the ground. I do this before I switch the fence on because the current flows better in damp or wet conditions.

I have used electric fencing to form paddocks for my free-range poultry; the squares on this netting are closer together than in the sheep netting, and the fence is taller. This netting not only keeps the poultry in but to some extent will keep foxes and dogs out. For cattle, a single strand of electrical wire will be sufficient. As a boy, I helped on farms and often had to move the electric fence for the dairy cows – giving them another area of grass to feed on. This is called strip grazing. I used to challenge myself to keep the moved fence in a dead straight line. Pigs need two or three strands of wire – they don't have a thick coat, so they really feel the shock. It may panic them at first, but they do soon learn.

Dry Stone Walls

You may take over fields or a farm with dry stone walls instead of hedges. You need to learn the basic principles of dry stone walling, which includes taking down (stripping out) a section of wall that is damaged and then rebuilding it. Courses are available and run by the Dry Stone Walling Association of Great Britain (DSWA); they have a network of local branches, and you will find most courses last for 2 days.

Gates

Metal gate posts up in my first field with post and rail fence. This is a 3.658 metre gate – nowadays I would recommend a 4.26 metre gate.

On a farm, metal gates are a good option. They are less expensive than wooden gates but probably don't look as attractive. You can always put a wooden gate in your farm entrance. With today's modern tractors and wide implements, you really need a gate that is 4.26 metres (14 feet) wide. As well as your metal gate, you need to purchase the metal posts, a hanging post and a slamming post. These are large metal posts about 2 metres long with a diameter of about 114 mm. The hanging post will come complete with pins onto which fit the eye bolts on the gate. The slamming post will have a latch or a slot. It will have a slot if the gate is fitted with a bolt.

Measure where your gate is going to go, and use string to get the gate in line with your two posts. I would lay your gate and both posts on the ground, and then you can see how much of the posts need to be buried in the ground. These posts need to be concreted into the ground, so when you dig out the holes, they don't want to be too wide, but leave enough room to tamper your concrete. The holes need to be 0.76–0.91 metres deep. Use a spirit level to make sure your posts are upright, and, once concreted in, leave them for at least 24 hours before hanging your gate. I like to see some post and rail fencing either side of the gate, which makes the entrance stronger than using just sheep netting. It also helps when moving livestock through the gate because they will see a secure barrier either side.

Wooden gates are a more expensive option, hardwood being more expensive than softwood. You need wooden hanging posts and slamming posts to match. The gate and posts should be treated with a wood stain preservative.

My new entrance gate, complete with brick pillars.

You can buy really smart timber entrance gates with the added option of them being electrically operated.

Chapter 5

Getting the Land into Shape

You may be lucky enough to have bought a field that is growing good grass or an arable crop that is safely fenced with neat hedges, with water, electricity, excellent buildings and possibly even a dwelling. The first field I bought in 1989 had none of these luxuries, but I hoped that one day it would have the lot and be a successful business.

My first field with weeds cut and taken away.

The previous owners of my field agreed to employ contractors to cut the weeds and cart them away at no cost to me, which was a great help. The hedges were cut by one of our local contractors, and fencing posts were knocked in ready for the wire.

Contractors agreed to come and spray my field with weed-killer – spraying almost 8 acres is too large an area for a knapsack sprayer, and so a tractor and sprayer are needed or a self-propelled sprayer. I used 'Roundup' – a systemic weed-killer. This will kill most weeds including grasses, bindweed, horsetail and nettles. The leaves absorb the weed-killer, which enters the cell sap and is carried to the growing points of the roots and shoots, thus killing the plant from within – even killing the roots. 'Roundup' is deactivated when it is in contact with the soil, and once it has dried on the leaves of the target plants, it is safe to walk through the field for you and your pets. After spraying my first field with Roundup, it killed most of the weeds but did not kill all the docks. Before sowing with grass seed, my students and I dug these out using a spade. Docks can be very difficult to get rid of.

Your new field will need testing for plant nutrients, including nitrogen, phosphorus and potassium. A specialist will come and do this for you, or you can buy a kit and do it yourself. Nitrogen is needed for leaf growth and yield, phosphorus encourages root growth and helps crops to mature, and potassium is needed for a healthy plant and the storage of starch and sugar.

The optimum pH for farm crops is between 6.5 and 7.0. Grassland, wheat and barley like more lime than potatoes and oats.

Take a good look at your ditches, if you have any. These are best cleaned out before you prepare your seed bed. If you wait until your seeds germinate, you will make a mess with the machine clearing out the ditch, and of course the rubbish and soil removed from the ditch will be spread on the land, and this will also make a mess. You must dig the ditch deeper than the level of the drain outfalls that run into it. Check your deeds to make sure the ditch belongs to you – you may need to enter your neighbour's land to clean out your ditch. After cleaning out, you will need a guard fence to stop livestock falling into the ditch.

If your land is wet or has water lying on the surface and perhaps rushes growing in areas, it will need draining. You will need a specialist firm for this if you are going to pipe the water away; however, you may be able to mole-drain the land if you have a clay soil with few stones. The mole plough is pulled through the soil, creating long tube-shaped spaces in the clay sub soil. These channels will deteriorate usually within 5–10 years; plastic drainage pipes will obviously last much longer. Luckily I have not had to drain any of my fields. By this time, you will have decided if you are going to grow grass or arable crops. The chances are if it is your first field, you will be growing grass, and this is certainly what I would advise. Before you sow your grass seed, you may want to mark out your farm drive and an area where you hope to put your farm buildings and even a garden. I just went ahead and grassed the whole field and then cut out the areas in the turf at a later date, but this is a waste of grass seed.

Tractors

If it is your first field, and you only have a few acres, you are best to keep any existing capital to develop your site and not spend it on a tractor and implements. You will therefore need contractors to do your work. When and if you purchase a few fields, you can then justify purchasing a tractor and some implements. Tractors come in all sizes – small ride-on tractors built for horticulture are too small for a 3- or 4-acre field. They would pull only very small implements with narrow working widths, and so you would have to go up and down the field too many times. It could be tempting to buy a vintage tractor perhaps a Fordson model N manufactured in the 1930s and 1940s, but it would not have a three-point linkage system and so could only pull trailed implements. This type of tractor is best left to the enthusiasts. A friend of mine was an enthusiast and owned a number of vintage tractors. He collected many scale models and books, but he has recently sold the lot, and he now tells me he is an ex-tractor fan!

Tractors manufactured in the 1960s often look very good, but they are not powerful enough these days and come with few luxuries compared with a modern tractor. I would suggest buying a modern second-hand tractor that would manage a reasonable-sized implement, be comfortable to ride in and of course have a cab. Hundreds of tractors will be for sale advertised on the internet and in farming magazines. New and second-hand ones can be very expensive indeed, or they can be purchased reasonably – prices will depend on the year of manufacture, the number of horsepower, the number of hours it has on the clock, the make and model, and the condition. Modern tractors now come with air conditioning and a lot of on-board technology, and you may now have satellite navigation. The tractor is often the farmer's pride and joy, and many would prefer this to a new car. An alternative to buying a tractor is to lease one. The teacher in our local primary school recently took a group of seven-year-olds on a farm visit. The next day, she asked the class 'What noises did we hear on the farm yesterday?' The replies were 'Moo,' 'Baa,' 'Oink, oink,' 'Cock a doodle doo' and 'Get off that ★★★★ing tractor!'

Contractors carry out all the tractor work on this farm. They cut 4.38 hectares of hay each year, and the mowing takes about an hour, which is quicker than mowing our lawns. It would take longer with a smaller tractor and mower. Turning the hay is again completed quickly with a large machine, and baling (which is small conventional bales), loading onto trailers and stacking in the barn takes about 6 hours.

If I had a small tractor and small implements and trailers, it would take much longer. There is therefore more chance of rain spoiling the crop. It would be less expensive to do the job yourself, but even with contractors making the hay, you can usually sell it for more than it cost to make. The decision is yours – you can buy a tractor and implements, employ contractors or perhaps do both.

Preparing a Seed Bed

You need to prepare a seed bed and sow grass seed on your new field.

Pig-tail spring tine harrow at work.

In 1989, contractors didn't plough my land but went straight in the field with a pig-tail spring harrow. Spring tine harrows or cultivators are usually used to loosen the soil and break down the lumps after it has been ploughed. This implement consists of rows of flexible iron tines fixed on a frame. It is mounted on the tractor's three-point linkage system. The pig-tail cultivator is very strong, and it did the job. It was a less expensive option than ploughing, but I really do prefer to see the field ploughed.

Five-furrow reversible plough.

The plough cuts and turns furrow slices burying the existing turf, stubble, weeds and surface soil bringing fresh soil to the surface and letting air in. Ploughs were originally pulled by humans, then by oxen, horses and finally tractors. The first ploughs were called trailed ploughs, and these are still used by enthusiasts who bring out horses and vintage tractors to do their ploughing and also compete in ploughing matches up and down the country. Today modern ploughs are mounted on the back of the tractor's three-point linkage and nowadays are usually reversible ploughs. These are fitted with both right- and left-hand bodies mounted back to back. If you plough first with the left bodies, you turn round at the end of the field and then plough back with the right bodies. This allows ploughing to be done avoiding ridges and furrows, which is what you get with just right-handed

– there is less marking out and less idle running. If you receive payment from the Basic Payment Scheme, you must not plough right up to the hedge – you need to leave a strip all around the field 2 metres from the centre of the hedge.

In recent years, there have been big changes in the preparation of a seed bed on arable farms. Instead of ploughing and harrowing, a single-pass cultivator can be used. These cultivators come in various combinations of discs, deep legs, working tines, presses, rolls and levelling boards, and different models are available for different working conditions. They prepare a seed bed in one pass, and you are then ready to sow the seed. Contractors have used one on my stubble, but we have found that it needs a power harrow to get a finer tilth after the single pass with the cultivator. Nowadays, really big machines are available with a working width of up to 10 metres. These 10-metre machines would need a tractor with immense horsepower, which is obviously for the contractors or farmers with a very large acreage. These cultivators don't give such a neat finish as ploughing, as you can see straw and stubble on the surface, whereas ploughing buries everything. However, it certainly does the job quicker than traditional ploughing and harrowing, and it has no ill effects on crop yields on this farm.

Single-pass cultivator on the arable field.

Single-pass cultivator preparing a seed bed for wheat.

Power harrows have rotary blades with a packer or crumbler roller at the back all driven by the tractor and so don't rely on the forward motion of the tractor. They turn over the soil, break up the lumps and smooth out the surface of the soil.

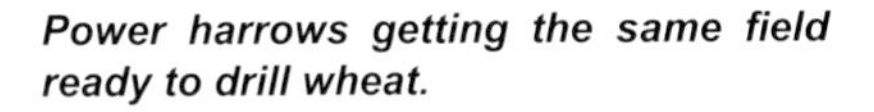

Power harrows getting the same field ready to drill wheat.

Rolls

Cambridge rolls leaving the surface in small ridges.

Flat rolls.

The Cambridge roller is made up of separate cast iron rings and leaves the surface of the soil in small ridges. It is ideal to use just before you sow grass seed. It will also break up lumps of soil and push stones down to prevent damage to machinery at harvest time.

Nowadays, many are hydraulic folding rollers with strong breaker rings, which fold up for transport on the road and for getting through gateways. When unfolded, they cover about 10 metres or more.

When I worked part time on Lord and Lady de Clifford's farms in the late 1960s, their roller was a simple trailed one that had a working width of 8 feet. One of their employees, Thomas, was sent off with a Fordson Dexta tractor and roller to roll the seed bed ready for grass seed to be sown. He set off around the field and, with a puff on his pipe, went round again. Unfortunately, he didn't look behind at the job he was doing, and each time he went around the field, a cast iron ring dropped off the roller, so his 8 feet working width was gradually reduced to 3 feet. Lord de Clifford spotted this and shouted very loudly, 'Thomas, you silly fool!'

The flat roller consists of a metal cylinder or cylinders. It will break up lumps of soil and make the ground smooth. It is also used to roll and flatten grass land.

Fertiliser Spreader

As well as spreading fertiliser, the spinning-disc fertiliser spreader is a popular method to sow grass seed. The application rate is easily adjusted, and the grass seed or fertiliser is fed onto a spinning disc, which throws the seed on the field.

Another method of sowing grass seed is a spring tine grass harrow and seed broadcaster combined. The harrow has a plastic hopper attached to the top, and fans blow the grass seed down tubes onto a spreading deflector plate, which spreads the seed onto the field. The harrows then cover up the seeds behind.

Zig Zag Harrows

This is a simple set of harrows consisting of a frame with a tine fitted where the frame members cross. Some farmers drag a zig zag harrow behind the fertiliser spreader when sowing the grass seeds to harrow the seeds into the ground. A heavy harrow will cover the seeds too much, and so a light seed harrow is needed. It is not essential to harrow your grass seed in – you can just roll after sowing.

Sowing Grass Seed

Don't go out and just buy a bag of grass seed. You need to find the right seed for the right job. First, you need to decide how long you are going to keep your grass. If it is sown and grown and ploughed in after 1 or 2 years, it is termed a short ley. After 3 or 4 years, it is a medium ley, and after 5 years plus, is a long ley. After this, it is termed permanent pasture, which can last for many, many years. A short ley is usually made up of Italian rye grass with perhaps different varieties in the bag. This is very quick-growing and high-yielding but with poorer growth in the second year, and then it deteriorates. It is cut perhaps three, four or even five times a year for silage. Medium leys could again be just for silage and would therefore be mainly Italian rye grass. Long leys and permanent pasture may have more of a mixture of grasses – perennial rye grass, hybrid rye grass and perhaps timothy, cocksfoot and clover – and so can be used for cutting and grazing. Perennial rye grass is one of the commonest species of grass and will last for many years; it is extremely hard-wearing. Hybrid rye grass is a cross of Italian and perennial rye grass, and has a mixture of both qualities. Timothy is a late-growing grass. It is very palatable and stays open all winter withstanding cold and wet conditions. Cocksfoot has deep

A spinning-disc fertiliser spreader sowing my grass seed with a zig zag harrow following.

roots and will survive a drought. It is not as palatable as the other species of grass. White clover is very palatable indeed – it is a legume that fixes nitrogen into the soil. Grass seed is sold by agricultural merchants in bags ready to sow. You don't mix the species of grasses yourself; it is all done for you, and it will say on the label what the species are. Depending on the mixture of grasses, grass seed is usually sown at the rate of 30–35 kg/hectare. The seed likes a pH of about 6.5. It can be sown in the spring, but if it comes a dry spell after sowing, the seed could be very slow to germinate or could germinate and then die from lack of moisture. I prefer to sow, and would advise to sow, in the autumn – August and September – when the chances are there will be more rain, and you should still have decent soil and air temperatures at that time of year. Grass seed can be drilled or broadcast – seed drills may get the seed too deep, and so broadcasting is best. The seeds should only be covered with 1 cm of soil.

Grass seed has germinated and is growing.

To achieve a seed bed, in my opinion, it is best to plough, power-harrow then Cambridge-roll (the roller being very important, as it gives a firm seed bed), sow your seed and fertiliser, and possibly lightly harrow with a zig zag harrow. Again, rolling is very important – roll with a Cambridge roller after sowing and then in the other direction with a flat roller. An old farmer once said to me that it should be so firm that you should be able to ride your bicycle over the field. If you can see the wheel marks it will want flat rolling again.

The grass can be grazed with sheep after about 8 weeks when the grass will be about 10 cm high. The sheep will do a world of good, biting off the grass, and then it will tiller – sending up more shoots. The sheep will also eat the weeds. They should only be allowed to graze lightly, and if it comes a lot of wet weather, take them off the field, as they will damage the seedlings. Watch out for leather jackets and slugs, and you may still have to spray for weeds.

Sheep and lambs grazing the new grass.

Chapter 6

Farm Buildings

A farming couple I know very well built their new calf building far too close to their house which attracted flies into their house during the summer. 'I'm fed up of these flies in the house' she yelled. Her husband was swatting them in the kitchen. 'I've just killed five,' he reported, and he went on to say, 'three males and two females.' 'How do you know what sex they are?' she asked. 'Three were on the beer, and two were on the telephone!' was his reply.

I do like to see an old-fashioned farmyard with traditional farm buildings made of brick with tiled roofs. In this area, farms were built about 250 years ago with a good strong house and the farm buildings often making up a square or rectangle with the house as part of one of the sides. The other sides were made up of loose boxes; each

Traditional farmyard.

one could house two or three beef cattle or a few calves, stables for heavy horses, a lovely long cow shed with standings for about 20 cows to be tied up to be milked, and an adjoining dairy with milk cooler and churns. A brick barn would be a taller building than the rest. It would be used to store sacks of grain and would have a loft above. There would be two or three brick pig sties – little brick huts with a small open brick yard in front of them.

The rick yard would be separate – situated at the back of the cowshed or at the side where hay and straw would be stacked outside in ricks. Cart sheds would be nearby, and as the years passed, these would be used to keep the first tractors. The garden and orchard would be at the rear of the house with a hen pen in the orchard for about 24 hens.

My Uncle Ben and Aunty Doll lived in the cottage at Cliff House, and my uncle managed the farm for about 30 years. This farm was more unusual in that it had stables and brick buildings for calves and cattle but not in a square pattern. They then moved to Home Farm in the early 1960s, and the buildings there were traditional, but their last move on the estate was to Overfield Farm, which was wonderfully traditional – sadly it has been sold and made into barn conversions.

When I left school, I worked at Harvey's Callendar Farm (two ells in Callendar for some reason but nobody knows why – I think the original farmer couldn't spell – Callendar Farm and neighbouring Callendar Grove made up 365 acres). Callendar Farm had traditional buildings along with a very large Dutch barn. After the war, Dutch barns were built to keep hay and straw dry, and as the winter progressed, and the hay and straw were used up, they would leave one or two bays empty. If you built some walls out of straw, it provided a good lambing shelter. Here at Oak Tree Farm, we have no traditional buildings.

During the 1960s, instead of cows being milked cows in cow sheds, they were starting to be milked in parlours more and more, and the cows housed in covered yards bedded with straw. Many years ago, they were, of course, milked by hand. Later, milking machines milked the milk into a bucket, which was tipped into the churn. Then, the pipe line came where the milk flowed into the glass pipe line and was pumped around the shed to the dairy where it was cooled and then went into the churns and later a bulk tank. Things certainly changed with the introduction of milking parlours. I haven't milked cows by hand, but I have milked with bucket machines and a pipe line and in a herring bone parlour.

Many cows were then housed in cubicles in the 1970s and silage clamps with concrete walls constructed. Beef cattle were being housed in larger groups in covered yards, and by the 1980s many sheep were being housed in the winter, which got them off the fields to rest the land.

Farms have got larger with farmers buying up neighbouring farms and hedges being removed in the 1960s to make larger fields. In the dairy sector, cow numbers have increased from 20–30 cows to probably over 100, with many herds consisting of 250 cows or many more.

Before the war, the farmer would be content with two or three sows rearing the piglets up to pork or bacon weight, and nowadays two or three sows would be classed as a hobby. Specialist pig farms now house hundreds or even thousands of sows, taking the progeny up to pork and bacon.

Tractors, combines and other implements are now enormous with bigger fields, higher yields, fertiliser and spray use – larger grain stores and potato and vegetable stores have had to be built.

In the poultry sector, again before the war, the farmer's wife kept hens to lay eggs for their own consumption or to sell a few for 'pin money,' being content to collect the eggs in a basket from a couple of dozen hens and rearing a few cockerels for the table. A popular number for free-range laying hens is now 32,000, and as for the broiler chickens, my friend James keeps 42,000 in one building.

Farms are no longer very mixed – there are exceptions, but many are specialist units with perhaps 250 dairy cows or thousands of pigs or perhaps 5000 acres of arable land.

So, with these specialist units and massive buildings, the days of the traditional farm buildings seem to be numbered. The biggest disadvantage is that they have small stable doors, which means that they are difficult to clean out – cleaning is usually done with a muck fork, digging the muck out and throwing it onto a trailer or muck spreader, but it can be done with a small machine called a bob cat. Another disadvantage is that these stable-like buildings are just not large enough. Some are rented out to girls with ponies, but sadly many of these buildings now remain unused. A few tiles may come off the roof and then a few more, and then the owner will say it's not worth getting repaired, and so it becomes derelict. The answer could be barn conversions. The owner doesn't have to sell them; they can be converted and rented out for homes, offices, storage facilities, workshops and factories. This will bring in a good income for the farmer whether they be rented or sold, which is surely better than seeing them derelict.

However, I call it asset stripping; I hate to see barn conversions. Your farmyard is no longer a farmyard. If these people had previously lived with their family on a nice farm with no neighbours, now look what they let themselves in for if they sell or rent out. Townies move in, thinking that they are country people but knowing nothing about the countryside, neighbours living very close – closer than on a council estate; listening to other people's arguments, noise, shouting, children's bikes to fall over in the yard, noisy dogs, cats carrying toxoplasmosis, cars parked all over the place, parties, awful barbecues (I hate barbecues!), people complaining about your cockerel crowing and waking them up, and vegetarians saying that you are cruel sending animals to slaughter – they might as well be living on a housing estate because their quiet idyllic life has gone forever. On the other hand, if a farmer is willing to sell you a barn conversion with a few acres of land, this can only be a good thing for you.

Barn conversions at Overfield Farm.

Planning Permission

Your new building would have to be necessary for your farm and meet certain conditions regarding, for example, the siting of the building and the materials to be used. I have summarised the main requirements that you need for planning permission, but you must familiarise yourself with all of the up-to-date information available at the Council offices.

You would need planning permission:

1. If your new building is to be built on a separate field to your main farm and this separate field is smaller than 1 hectare.
2. If you wish to build a dwelling.

3. If you wish to build a livestock building or a slurry and sewage store.
4. If the ground area is 465 square metres or larger – this rule is further complicated because this area must include the area of any other buildings that have been erected within the preceding 2 years and are within 90 metres of the proposed new building.
5. If the building is to be 3 metres or higher if it is within 3 kilometres of an aerodrome – this is unlikely, as most farms are not next to an aerodrome.
6. If the building is going to be higher than 12 metres (and not next to an aerodrome).
7. If the building is within 25 metres of a classified road or a metalled part of a trunk road.
8. If the building is within 400 metres of the boundary of land surrounding a dwelling or protected building.
9. If the land requiring excavations is connected with fish farming.
10. If you are building in a protected area such as a National Park.

You will need to visit the council offices with all of your plans and notes; then after consultation you need to fill in the forms and do things by the book, and hopefully you will get your building, but be aware if you are close to a village or have 'townie' neighbours, you could be subject to a public debate. Your opponents may raise issues with the council such as road safety or that they think it is a blot on the landscape, or your slurry or manure could cause a smell and encourage flies.

Sometimes these people can be very determined, and you may be the subject of a newspaper article or even television coverage, but this is unlikely. However, you may have to compromise, and I do realise I was lucky with my farm buildings, as I had no trouble from the council or anyone else. When planning where to put your building, you need to think about leaving a lot of space for tractors and trailers to turn around outside the building and get into the building. Also, many feed and other deliveries now arrive on articulated lorries, so you need more space than you probably think. Don't forget drainage. Hopefully water and electricity supplies will be close to hand, but the further away the services are, the more money it will cost to get them into the building. You should try to find a fairly level site, but this is not always possible. Plenty of hard core and concrete will be needed to bring up the levels. Cattle and sheep buildings that are open fronted, if possible, should face south or south-east. Don't forget your planning permission – you are unlikely to get permission if your new building is too close to other people's dwellings, or if there is a problem with access to it off a very busy road on a bend. Specialist firms will have standard farm buildings on their books, but they will also design buildings to fit in with what you have got or to extend existing buildings. They will help you with planning permission.

A small building is likely to cost more per square metre than a large building so think as 'big' as you can – an extension at a later date is likely to cost you more. You can purchase buildings in kit form and then get another local firm to erect them, or you can do it yourself with farm help or friends' help, but to do it yourself, your farm may suffer in that other jobs might not get completed. More expensively, you can get the specialist firm to supply and erect your building. Check whether a construction firm charges by the hour or to do the whole job, and some firms will charge you travelling time from their depot on top of labour once they get to you – that adds up to a lot of money – for example, four men sitting in a lorry for 1½ hours travelling to you and 1½ hours travelling back. Not only do they charge labour for sitting in the lorry, but diesel will be charged to you as well. Get plenty of quotes from different firms. Before starting work on your farm building, I suggest you go and look at some and talk to the farmer and farm workers so that they can point out anything that is badly designed. Steel portal frame buildings are the most popular buildings and are built to last. There are specialist firms that will sell you the steel-framed building and the roof, and also cladding, or you can fit your own cladding. Timber buildings can also be purchased for all livestock. They probably work out cheaper to buy than steel, but steel-framed buildings can be built wider than those made of timber; a large span makes cleaning out easy, and there is no or very little maintenance. Wooden-framed buildings tend to be smaller, and wooden posts erected in the centre of the building to hold the roof up make cleaning out more difficult. Maintenance costs can be quite high if the timber needs to be treated with preservatives. Wooden posts do sometimes break or rot off at the bottom. Wooden buildings, however, may integrate with the landscape better than steel. Concrete floors are essential for ease of cleaning, as soil floors are a magnet for bacteria; also, they will sink, and every time you clean them out, you will take some soil with you.

Some buildings are seasonal and are used for only part of the year, for example, a lambing shed, but if this building could be utilised for the other parts of the year, this could be beneficial – to house young cattle for example. Recently, a Birmingham teacher asked one of her 14-year-old students to name the four seasons. The answer given was salt, pepper, mustard and vinegar.

Livestock Housing

Buildings provide shelter for animals, a safe and comfortable place to stay, somewhere pleasant for you to work and, above all, a place where the animals will 'do well' and harvested crops and machinery can be kept dry and safe. They are expensive but, in the long run, are a good investment, adding value to your farm.

If you are a tenant farmer, the landlord may agree to pay for some new buildings or agree to let you build some at your own expense. However, if you ever come to leave the farm, you may be able to take the building with you but not the concrete floor, water pipes, electrics or drainage pipes.

Fresh air is essential; stale air and the build-up of ammonia will encourage respiratory diseases, so the movement of air is important – fresh air needs to come in to the building, and stale air and heat need to go out. Draughts at the same height as the animals are not good for them and could give them wind chill.

Cattle Buildings

You need a big strong shed for growing cattle and adult cattle. It needs to be tall enough to get machines in to clean it out – at least 4 metres to eaves height – and it needs strong galvanised gates. A concrete floor is essential, and then you can have timber buildings, but for large cattle steel-framed buildings are best. Foundations will be below ground level and strong enough to hold the walls. The bottom half or two-thirds of the building will be concrete blocks, with concrete poured inside them, or pre-stressed concrete panels – these can also be used inside the building and can be dismantled and moved to change the layout. The top half or third of the walls will be made up of Yorkshire boarding – this is space boarding – boards nailed or screwed on with gaps in-between for ventilation. Alternatively, box profile steel sheets can

Cattle building with Yorkshire boarding.

be used to clad some of the walling – usually 0.5 mm gauge with ribbed air vents incorporated. Roofs are best sheeted with galvanised steel box profile sheets. These have a plastic-coated finish or a polyester finish in 0.7 mm gauge or 0.5 mm gauge (0.7 mm is usually used on cattle buildings) with a cover width of 1 metre. These are available in a wide range of colours. Wide eaves that overhang outside the animals' area will provide added protection against the weather. An anti-condensation layer can be stuck to the underside of the sheet; this is done at the factory.

Fibre cement is an alternative for roofing. This usually comes in a thickness of 1 mm. It looks very similar to asbestos sheets. A rooflight is a clear sheet that lets light into the building. At the top of the roof, an open ridge will provide outlet ventilation, or different types of open ridge vents can be fitted. Dairy cattle are housed in covered yards or cubicles, and beef cattle in covered yards.

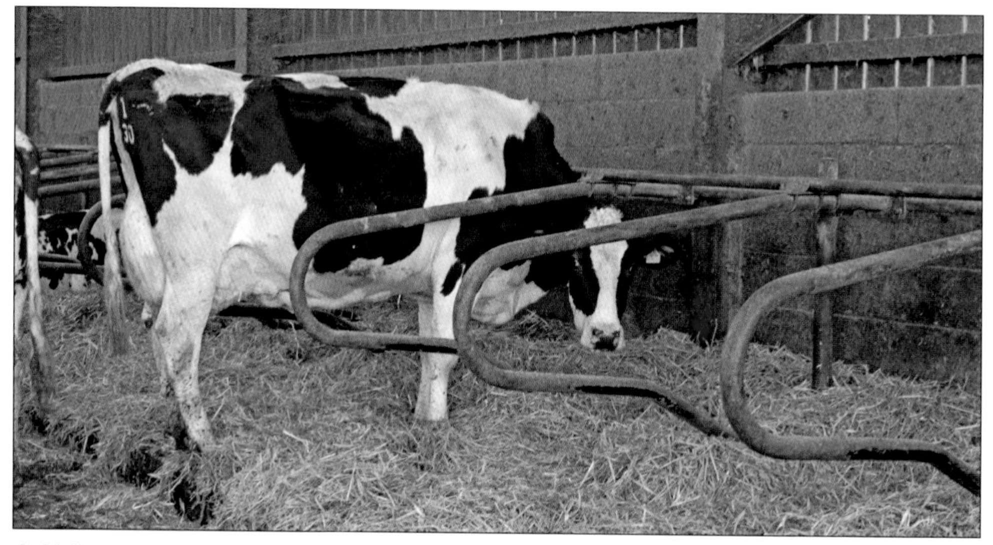

Cubicles.

Storm water should be clean with no sewage effluents, and then the clean water is discharged into a soakaway or into a ditch – these days through plastic drainage pipes. Fitting a septic tank will be the way to treat sewage, where bacteria digest the waste products, and on large dairy farms slurry is stored in large slurry lagoons.

Calves also need a well-ventilated, draught-free house. Newborn calves like to be warmer than dairy cows and large beef cattle, but if the air becomes stale you are asking for trouble with disease. They should be able to lie down comfortably and see other animals in pens that join. Individual pens inside the building need to be 1.5 metres × 0.75 metres, but larger if possible – about 1.8 metres × 1.0 metres. They are put into groups at no later than 8 weeks of age. Again, in these groups, ventilation is important:

too many calves in together will mean there is not enough air space, and no more than 30 calves should share an air space.

With loose housing of cattle, they should all be able to lie down at the same time and get up without difficulty. Growing beef cattle and dairy heifers want more room as they grow. A 200-kg animal would need about 3 square metres in total of floor space, and a large 600-kg animal would need a total area of about 6.8 square metres. You can check with DEFRA for minimum requirements, but if you crowd them in too thickly, you are increasing the chance of respiratory diseases.

On a dairy farm, most cows and some heifers are housed in cubicles. There should be one cubicle per cow, unless there is plenty of loose housing adjacent and available to the cattle. Each 600-kg dairy cow would need a solid floor area, plus loafing and feeding areas, totalling about 6.80 square metres.

Straw yards and cubicle buildings should have a concrete feed passage, and loafing areas to prevent feet from becoming overgrown. The feed passage wants to be wide enough for the cattle to be able to stand in line and eat but with enough room for stock to move behind them.

Calf building – this is an older building, which could be purchased for much less money than a new building.

Sheep Buildings

Sheep buildings can be constructed in the same way as cattle buildings, and I like to see one side of the building open and fenced off with feed barriers or gates. Sheep don't like to be hot – an open ridge will also increase air flow.

A less expensive way of constructing a sheep building is to build a pole barn. This is constructed from telegraph poles sunk into the ground, with trusses made out of timber and a box profile roof attached, with some cladding on the windward sides.

Sheep building.

Inside the sheep building, 90-kg ewes would want a floor space of 1.2–1.4 square metres per ewe, but lambs at foot and still being kept indoors would need 2.0–2.2 square metres of floor area; however, less space is needed if the ewes are winter-shorn.

Pig Buildings

Pigs have only short hair and feel the cold and heat more than other farm livestock. Many pig buildings are environmentally controlled with air vents and fans, while other buildings rely totally on natural ventilation.

Building suitable for weaners, growers or fatteners.

Pigs can suffer from heat stress, and so pig buildings should be insulated to minimise heat in the summer and keep the building warmer in the winter. All buildings need really good ventilation, providing fresh air in and stale air out. The speed of the air and the number of pigs in the building will affect the temperature.

The farrowing house needs to be warmer than other pig

buildings; it should be totally enclosed and constructed of a combination of brick, concrete blocks, timber and box profile sheets. Sows and piglets would remain in these buildings until weaning.

Buildings for older pigs can be constructed with sheeted gates on one side and then an open gap above these gates. This gap should be closed with sheeting in very cold weather.

These buildings can have a closed-in box at the back only just over the height of the pigs similar to large kennels that pigs can enter to go to sleep and keep warm. The outside structure will probably be steel-framed with concrete panels or blocks on the bottom half and Yorkshire boarding on the top half or box profile sheeting.

Stocking rates with pigs is common sense – they must be able to lie down comfortably and not be overcrowded. If they are overcrowded, respiratory diseases may occur. On straw finishing systems, a 100-kg pig requires 0.93 metres square, which gives it enough room without touching another pig. The general recommendation is 1.4 metres square per 95-kg live weight. The minimum space for sows is 2 metres square.

Inside a pig building showing the area at the back for warmth.

Poultry Houses

Traditionally, timber buildings have been used in the poultry industry for hens, pullets and broilers. These are made to specification by specialist firms, or you can buy the materials and build your own. Internal layouts will differ in individual buildings.

Wooden frames are constructed, and then buildings are cladded with timber. Inside, they can be lined with a smooth-faced white steel sheet. Ventilation systems will vary, but adjustable air vents on the sides bring fresh air into the building, and fans in the roof take the stale air out.

Box profile sheets are used for the roof with a bottom sheet then insulation material on top of it and a top profile sheet. Three layers will insulate the building and stop condensation. An alternative that is more expensive is to use composite panels – these consist of a top profile sheet, a layer of insulation and the bottom layer all put together as one at the factory. Most buildings have no windows with a controlled environment but for high-welfare buildings for free-range and organic birds, double-glazed windows and shutters are available.

More and more new buildings for poultry are steel-framed with galvanised steel profile cladding and a roof with either a plastic or polyester finish or composite panels – the outside walls are usually dark green, but other colours are available. These buildings are more expensive than timber buildings – perhaps 30% more – but the advantage is that you can have wider spans with no posts in the centre of the building.

Laying birds are housed at a density of 9 hens per square metre; for pullets (Freedom Food standards), 14.5 birds per square metre; and for broilers, 38 kg per square metre.

Broiler house.

Ducks and turkeys can be kept in environmentally controlled houses with natural ventilation – geese go out in the grass fields in the daytime and so don't want a controlled environmental house. Ducks are housed at 25 kg per square metre. Turkeys are housed in broiler-type housing at 260 cm square per kilogram or in pole barns at 410 cm square per kilogram.

Hay and Straw Barns

I like to see hay and straw stacked neatly in a building and not turning black outside with the weather. First-generation farmers can construct their own pole barns using telegraph poles and galvanised steel profile roofing with some side cladding, or, if finances are available, specialist barns can be sourced. I am amazed at how many established farmers leave their straw outside to get wet. I won't leave one bale out.

Hay and straw barn.

Tractor and Implement Buildings and Workshops

These can be similar to the hay and straw buildings with perhaps a section bricked up with windows for a workshop, which can be locked.

Farm workshop.

Grain Stores and Potato Stores

These are for the specialist arable farmer, and it will only be a few first-generation farmers that will have enough acreage for these buildings.

Grain store.

Chapter 7

Developing the Business at Oak Tree Farm and Constructing Farm Buildings

You may be lucky in purchasing your farm or field, or fields complete with farm buildings. I wasn't that lucky. I needed to turn my first field into an agricultural business, and so after clearing the site, I certainly needed buildings. Pigs or poultry were the only real options to make a living because of the small size of my land. It wouldn't carry a dairy herd, and I could only keep a few beef cattle or sheep, which would not be enough to make a living, but I was almost sure I could make a living from pigs or poultry.

You need big strong buildings for pigs unless you have them outdoors. Pigs produce a lot of muck and are hard work, and so poultry was the way forward for me, with some sheep. With sheep, you can get away without buildings if you buy in-store lambs and then sell them as finished lambs in the autumn and winter, but for a serious breeding flock, you need buildings. These buildings are not just for shelter; by wintering the sheep inside, it gets them off the field, which gives the grass a chance to recover and grow, preventing the field from becoming poached and a sludgy mess. ('Poached' is when the land becomes packed and trodden, leaving areas where the grass struggles to recover.)

It is better for me to carry out lambing during February and March inside a building rather than outside in a cold, wet field. I keep the ewes out for as long as possible, but on this clay land poaching will occur, and so if it gets very wet under foot, they come in, usually some time in January.

Free-range laying hens need a building that is environmentally controlled with lights and fans, which is expensive. It is no good building this if you don't have a market for your eggs.

My old farm boss, Mo, didn't bother with 'fancy farm buildings,' especially to store his machinery; he would forget where he had parked it, and then in the middle of the summer, we would have to search through the stinging nettles to find the mower and hay turner.

Poultry Building

My first agricultural building was a poultry house for 3000 laying hens. In those days, I didn't need any planning permission or building regulations. To start, I looked at second-hand timber poultry houses for sale in agricultural magazines. At the same time I found an advertisement from John Bowler Agricultural Ltd – a franchise firm that would sell me a second-hand building, dismantle it, transport it and re-erect it on the farm, and then sell the pullets and agree to buy the eggs and end-of-lay birds.

My father and I visited John Bowler's office at Etwall in Derbyshire in the summer of 1990. We had a meeting in the office first, and we were shown what 'Bowlers' could do for us. We were also shown how much we needed to spend and how much profit we would hopefully make. The facts and figures looked good, and after coffee we were shown a 5000-bird unit, which was a large unit in those days. I talked it over with my father, and he encouraged me to sign up for 'Bowlers.' They would provide a second-hand shed and the birds, and take the eggs and the end-of-lay birds.

There were a number of second-hand buildings on the books, and a suitable building measured 48 metres long by 7.5 metres wide, 1.9 metres to the eaves and 4 metres to the top of the roof ridge. This building had housed pigs and was available for £2000. It would house 3000 birds for me on a free-range system.

I signed on the dotted line and was very excited. I was a landowner because I had previously purchased my field outright, and I owned some breeding ewes, lambs and rams. I had made some hay, but now this was a really big project with the option at a later date of hopefully buying more land, more sheep and more hens. I would no longer just be farming at my school; now, I felt like I was farming for real, spending some serious money to hopefully make more serious money, and to prove that I wasn't just playing at it, I needed to register for VAT.

I told John Bowler straight after I had signed that I dreamt of making a million pounds out of poultry like my farming neighbour. 'Did he achieve that?,' asked John. 'No,' I replied, 'he just dreamt of it!'

After signing, I travelled home, knowing my life would now change – free-range hens would become a big part of my life and would be a very big commitment. I would hopefully build up the business, and at a later date when the business became viable and I had plenty of work, I could live on site in a mobile home, and then a few years later perhaps a bungalow or a house might be possible.

John Bowler had got producers on his books that had gone down this route and were now living on site in a mobile home, house or bungalow. I would still need to keep teaching for a long time yet – which I didn't mind because I enjoyed my job – but the ultimate aim was to have a viable business, live on site and eventually give up the day job.

Two thousand pounds seems inexpensive for a poultry house. I bought the building without seeing it – I had just seen photographs. I wouldn't advise anyone to do this, but the building was in Reading, which was a long way off, and I trusted the Bowler team. However, this £2000 was only the start, and I was told that in order to get the building up and running, I would be looking at a total cost of approximately £30,000, which wouldn't include the purchase of 3000 point-of-lay pullets. It needed to sit on three rows of concrete blocks, and so I worked out how many I needed, ordered some spares and had them delivered before the building arrived. The land was dry, and so the lorry brought them right onto the field. The Bowler team dismantled the building and transported it on two lorries to my field. I was certainly not disappointed: the building looked almost new and was very clean indeed. It had been built in sections, each section 2.43 metres long (8 ft), and each section would bolt together with the two gable ends both in one section. The building was constructed of tongue-and-groove boards nailed onto a frame with an asbestos roof. Asbestos is now classed as a hazardous substance, and is no longer used in farm buildings. One section would be converted into the egg room, the next section was where the food bins would go, then a walk-through passage down the centre, and the rest would be for the hens.

The roof trusses were strong and looked amazing, they certainly gave the building some character; nowadays they wouldn't look out of place in a barn conversion. I decided to situate my building towards the back of the field away from the road but parallel to it, placing it only 13.70 metres from the hedge on the long side and 5 metres from the fence at the gable end where the entrance to the egg room would be.

After signing on the dotted line, work commenced about 6 weeks later. After measuring up the exact spot where it was to be positioned, the next task was to take out the foundations. I had agreed that most of the building would just have a soil floor; this was to save money, but I would advise a concrete floor for ease of cleaning. The areas that needed concrete were where the walls would go, a central egg-collecting passage, the egg room and a strong base for the bulk-feed bin to sit on.

They started with the foundations for the walls. I originally expected them to dig out trenches with the turf and topsoil taken out, hardcore placed in and then finally the concrete. However, this was not the case. I was told at the initial meeting that the walls would just sit on a concrete ring. To level the ground, some of the turf and topsoil was taken out on part of the field, but at one end the concrete would be laid straight on top of the grass.

Two shuttering boards laid parallel to one another were placed on top of the grass 25 cm apart and standing 30 cm high. These shuttering boards continued all the way around the perimeter of the building – forming a ring. I had ordered the concrete from a local firm, again the field was dry, and so it was easy for the ready-mix lorry to enter the field and deposit the concrete between the shuttering boards – I used a 4-to-1 mix, that is 4 parts sand and gravel to 1 part cement.

I was not really convinced. It is a clay soil, but I thought that with the weight of the shed walls and roof on this concrete ring beam, it would surely sink, but it didn't, and it is still in place today.

A little topsoil was dug out for the central passage – a passage that would run the whole length of the shed where the birds were housed. This passage was for us to walk down and collect the eggs from the roll-away nesting boxes. Hardcore was placed in and then the concrete, which was 15 cm deep this time.

The foundations for the new poultry building are under way. The concrete base for the egg room is in the foreground, and the concrete ring beam is completed on the right.

The egg room was treated in the same way, but the base for the bulk feed bin needed to be stronger to support a bin that would hold 7.5 tonnes of feed, and so the depth of concrete this time was 30 cm. I asked the lorry driver if he had heard about the football match between Rugby Cement Works and Coventry Sand and Gravel Works. 'No' he replied. 'It was a good game,' I told him and then added 'What do you think the score was?' 'I've no idea.' '3–2 on aggregate' was my reply.

I found a local brick layer who was reasonably inexpensive. He was called Bob, but this was in the days before Bob the Builder. He laid three rows of concrete blocks on top of the concrete ring beam ready for the building to sit on. I provided him with the sand and cement, and he brought his own cement mixer; the building was starting to take shape.

The sections that made up the walls were bolted together, and amazingly, with the help of a large sledgehammer, everything fitted into place. It was then time to put on the roof trusses and purlins, and then the asbestos roof. The egg-collecting passage up the centre would divide the building into two, and so 1500 birds would go on each side.

Most of the walls have been fixed to the concrete blocks, and most of the roof trusses are in place.

The roof is now almost completed.

The timber and wire floors needed to be made on site. The birds needed to be housed on these floors about 1 metre above the floor of the shed. The floors would rest on turned-up railway sleepers. The idea is that the birds walk around on the wire floors, and then the droppings drop into the pit below. To make the floors, a wooden frame was constructed with strengthening timbers inside the frame, and then some very strong wire was stapled on. The floors were very large, measuring 2.4 metres by 2.4 metres after completion. I was amazed at how heavy they were. It took four men to pick them up comfortably and position them inside the building. It did occur to me that when it would be time to sell the birds these floors would want lifting out which would take another four men to do the job.

The birds, of course, would need nesting boxes, and these would be made from old battery cages. The cages came from a poultry farm that had stopped producing. They were delivered on site and were then converted into roll-away nesting boxes – these have sloping floors so that when the eggs are laid, they roll away out of the bird area into a collection tray which faces the central passage.

Four fan boxes were made to fit on the roof. The fans arrived and were fitted. Vent boxes were also constructed on site and fitted on the outside of the building. The vents could be opened or closed from the inside of the building – the combination of fans and vents meant that the building would be well ventilated, which is very important especially in hot weather. Pop holes were cut on the outside of the building so that the birds could get in and out, and then ladders were made so that the birds could get down to ground level and back up again.

Fan boxes have been fitted in the roof, vent boxes fitted on the sides and pop holes cut so that the birds can get into the field.

I bought a 7½ tonne bulk feed bin that was delivered and bolted to the concrete base. An auger was also purchased – this is a long plastic tube with a metal spiral inside that takes the food from the outside bulk feed bin into a bin inside the building. With 3000 birds, it is no good buying bagged feed because it is much more expensive purchasing it this way and, of course, you don't want to be handling bags of feed every day.

The next piece of equipment is a chain feeder, which takes the feed from the inside feed bin to the birds in the shed. A chain pulls the feed along in a metal trough, which runs around the shed, thus eliminating the need to lift, carry and pour feed out of bags. An electrician then came and connected up the lights, fans, auger and chain feeder to a generator. I had no mains electricity to start with. The lights, fans and the chain feeder were all put on time clocks so that they would work automatically. Bell water drinkers were fitted, which would give the birds a source of fresh drinking-water automatically. When a bird takes water from the drinker, the drinker becomes lighter in weight. Fresh water flows into the drinker, bringing it back to the original weight by replacing the water taken out. The water was connected to the mains water, which was metered and came from the neighbouring farm. It was not very good, as it trickled down in old lead pipes. I would have to wait a while for a mains electricity supply and a new mains-water supply. Electricity came first, about 3 months after the purchase of the pullets, and the water about sixteen months after this.

The Bowler team of four men worked 5 days a week for 5 weeks and finished the work except for fitting a door to the egg room. The following weekend, my father made a door – good old Dad!

A week after finishing the building, my new pullets arrived. We had created a controlled environment inside the building with birds up on floors out of the way of the muck. We

My father on the right with a young helper fitting the door.

Inside view of the building showing timber and wire floors, bell water drinkers in the centre and metal chain feeders on the left and right. On the far right, you can see the roll-away nesting boxes.

had lights, fans, automatic water drinkers, an automatic feeding system and all 3000 birds housed in the one building with access to the field in daylight hours. To collect the eggs, you walked up and down the central passage taking the eggs from the roll-away nesting boxes and placing them onto egg trays. My mother happened to mention to Andrew – one of my former pupils – that we could do with a trolley to put the trays in. The next morning, an old shopping trolley had been lifted over the road gate and left at the top of the farm drive!

I needed a farm drive, but to save costs in the early days I just dug out two trenches where the wheels of lorries and cars would go and filled these trenches up with hardcore. It wasn't very satisfactory, and if a driver strayed from this make-shift drive, he could get stuck, especially in wet weather. However, I had to wait a little while before it was improved. A few months later, I had the drive dug out, hardcore put in and then road planings put on the top, which improved access no end.

The building is complete.

Up until now, most free-range hens lived in small groups in small sheds with perhaps extra lighting – some farmers still operate this system. Often these sheds were mobile and were drawn around the field perhaps weekly; indeed these sheds still exist on some farms and are ideal for smaller flocks. Larger buildings like my new set-up were the way forward, and over the years this system has improved again and again. I was getting good at keeping free-range hens, and I was now ready to expand. I increased the bird numbers to 4000 birds. This meant that I had more man hours and also more profitability – two factors I needed before I could justify a mobile home.

I really needed 5000 birds. In 1994, I met a relation of my old farm boss who happened to say, out of the blue, 'Do you know anyone that would like to buy a second-hand poultry building?' I replied, 'Yes, me!,' and I went up to look at it the following weekend. I took my tape measure, and even at first glance it looked the same as the existing building but not as long. It had tongue-and-groove boards – almost identical to the existing building. I bought the building for £800. It was 22 metres long and just slightly wider than the poultry house I already had. I called Bowlers, and they agreed to come and put it up. It was such an excellent match, and it made my poultry unit 70 metres long with the new extension.

Again, a concrete ring beam was laid, but it had to be taller this time because the land sloped away. It rained and rained, and the building site was a sludge pit. We again had to lay three rows of concrete blocks, and then it was easy bolting the sides together. Because this building was slightly wider, Bowlers had to shorten the gable end and shorten and alter the pitch of the roof trusses. The asbestos roof matched very well. The end of the existing building was, of course, taken out.

A new second-hand extension has been purchased, and the walls and roof trusses are being fixed.

I bought some second-hand roll-away nesting boxes, which saved the bother of altering battery cages. Floors were made here on the farm. I purchased more lights, another chain feeder, water drinkers and two more fans, and at last the job was finished. It all fell into place nicely because at the time we were building, the birds had been sold and the shed cleaned and washed out, and so I could move the new flock of 5000 in.

When it was all finished, you couldn't see the join. From the outside, it looked as if it was originally one building. It was a first-class effort from Bowlers.

I had achieved another goal – 5000 birds – and it certainly meant more man hours and more profitability, which would help me when I was ready to apply for planning permission for a mobile home.

My father sorting eggs in the egg room.

Since 1994, I have made a few changes to the poultry house, some to make life easier and some because of changes in the rules and regulations. The first change was to cut the wood and wire floors in half to make it easier to lift them in and out. This took longer than you would imagine because you have to cut the timber and the wire, and then construct new ends. During 1998, contractors concreted the floor in the building. This was essential as the soil floor was impossible to clean, and every time we cleaned the shed out we took some soil out, too. The floor was now much lower than the field outside, and so rain water kept draining in and flooding it. A second larger bulk feed bin was purchased – with two bins, I could now buy 18-tonne loads of feed, which was cheaper than 6-tonne loads.

During 2006, new ruling from the Department for the Environment, Food and Rural Affairs (DEFRA) meant that all free-range hens had to have a scratch area in their building. I could have put this inside the existing building, but that would have meant reducing the number of birds I was permitted to keep. I made the decision to have a new building built onto one side of the poultry house. This would be in the form of a lean-to running almost the whole length of the shed on one side. We built this on the farm with help from local farmers and contractors, Brian and Ben (who also clean out the building with their caterpillar tractor) and my farm students. It was constructed from timber with steel box profile roofing sheets, and the inside of the roof was lined with plywood to stop condensation. Wood shavings on the floor completed the job. The birds can come and go using the pop holes in the original building to get in and out of

A new lean-to scratch area under construction.

the lean-to, and of course we made new pop holes so that they could go out into the field.

The new lean-to completed.

After the birds were sold at the end of lay during 2008, I had new automatic nesting boxes fitted. This wasn't a new rule from DEFRA, but it would make my life very much easier indeed. The boxes are very comfortable, and the eggs roll away much more easily than the old boxes that were converted from the battery cages; now they rolled onto an automatic egg collection belt. The eggs were much cleaner with this new system, and certainly there were fewer broken eggs. The birds are not disturbed because you are not walking up and down the passage with a trolley. The belt brings the eggs into the egg room where they are manually taken off the belt and placed into trays. The new nesting boxes cost £10,787, including the labour to fit them.

We had to rip out the old egg-collecting passage and the old nesting boxes – there was no need for a central passage. The new boxes were put where the passage had been – right down the centre of the building. Of course, they didn't take up exactly the same space as the passage, so the old floors wouldn't fit. This was an excellent time to replace the old floors, especially since a new ruling from DEFRA meant that you couldn't house birds on wire floors any more.

The new plastic slatted floors are again raised off the ground so that the droppings fall through into the pit below. They are more environmentally friendly, and they are warmer and kinder to the birds' feet. These floors are easy to handle and fit together like a jigsaw. They measure only 90 cm by 60 cm and are easy to jet-wash clean. They cost £6360 including the labour to fit them. I was able to sell the old floors, and my customer was able to alter them and make pheasant-rearing pens out of them.

I found that the bell water drinkers did waste water. If birds knocked them, they would tip and spill water, making the droppings in the pit below very wet. I had new automatic

All now very modern – with automatic nesting boxes on the far left then the red automatic water drinkers. The birds are standing on the plastic slatted floors and feeding from the chain feeder system.

nipple drinkers fitted in 2011, again connected to the mains-water supply giving the birds a constant source of fresh water. The new drinkers cost £2249 with materials and labour. The automatic nesting boxes, new floors and water drinkers were all supplied and fitted by ELM Ltd – Equipment for Livestock Management from Wellingborough, Northamptonshire.

During 2012, again new ruling meant that you can only stock inside the building at a rate of 9 birds per square metre. This was reduced from 11.7 birds per square metre. However, this didn't affect my system, as the building and scratch area were large enough.

Pop holes are now larger, which encourages the birds to go out more. Many units have automated pop holes, which open up in the morning and close at night. When you get the new birds, for the first few weeks many birds don't go in at night. You have to catch them and put them through the pop holes, and even months later you will often get a few strays out at nights that need catching. If I left them out, foxes would kill them, so I prefer to walk around and look for birds in the evening rather than have automated pop holes.

During 2013, we put a new roof on the whole of the building. I purchased all the materials, and then Lindley Hall Farms with Colin in charge did an excellent job fitting new purlins and then fitting a bottom galvanised steel profile sheet (polyester-coated),

New roof.

then insulation on top and finally a top sheet of galvanised steel profile (polyester-coated). The whole project cost just over £22,000.

Sheep Building

I had kept my flock of pedigree Kerry Hill sheep on grass keep and took them into school to lamb them there. Educationally, my flock was a great benefit to my students, but I wanted to increase numbers and keep them on my new land. I had built the sheds at school first making a wooden frame and then nailing on tongue-and-groove boarding. The end section was then bolted together. The largest shed at school measured 15.4 metres by 7.5 metres. It housed not only sheep but goats, calves and rabbits.

I knew how to make a building but a sheep building on a farm needed to be much larger and much taller, and it needed to look professional. It would also be used to store hay and straw. John Bowler's firm agreed to come and build it; not only would it be a useful building, but having a well-built building on site would be favourable for me when it came to getting permission for a mobile home. To lamb a flock of sheep on site, I really needed to be here so that I could get up in the night and check the ewes.

It was the autumn of 1993; the new building would consist of four bays each measuring 4.6 metres by 9.2 metres and the total dimension of the building 18.40 metres by 9.2 metres and 3.6 metres to the eaves.

Ideally, I would have dug out the soil and put in hardcore and then a large concrete base on top of the hardcore. However, at this stage, to save costs Bowler's just dug out the foundations where the posts would go to hold up the roof; this included six posts inside the building and 14 posts around the perimeter (125 mm by 75 mm). They also constructed a concrete ring beam where the walls would stand.

The building was constructed in sections just as I had done at school; however, ventilation in this new building was improved. On the two short sides and the one long side (the back), each section was made up of exterior ply, which was fixed to the bottom and stood 1.4 metres high. On top of this was Yorkshire boarding, which is good for ventilation. The front of the shed was open at the top with wooden doors made out of exterior ply wood at the bottom. The roof was second-hand asbestos sheets fixed to 175 mm by 125 mm rafters. Ventilation is very important in a sheep building owing to the risk of pneumonia in a closed environment. The total cost of materials and labour was £1000 per bay, £4000 in total, but this did not include bringing water and electricity inside.

Walls and roof being fitted to the new sheep building.

Sheep building with three bays for sheep and one bay storing hay and straw.

One of my students, Andrew, constructed some timber and wire hay racks, which finished the building off. I lambed 51 Kerry Hill ewes in the building during the spring of 1994 and have lambed ewes in there ever since.

I was so pleased with the building that I asked John Bowler's team to come back and construct another bay on the far end, and this was less than 12 months after the first build. I now had five bays. The building now measured 23 metres by 9.2 metres, and another £1000 had been spent, which was money saved from my teacher's salary.

I was so pleased with the new sheep building that I am having another bay built.

During 2001, I improved and updated the sheep shed. Farmers and contractors, Brian and Ben, concreted the floors, and new gates were fitted on the front of the

The floors have now been completed in the new sheep building and new hay racks and troughs fitted.

building. Inside, some smart new hay racks and troughs replaced the old ones, making it a first-class building. It has also been used to house the show sheep and store hay, straw and sheep feed. It would be very difficult to run the farm without this building.

One of my friends, Harry, spent far more money than me on a very large steel-framed building with all the extras. After spending this vast amount of money he had a slight problem during the first week of housing the ewes. Just at the wrong moment, there was a power cut. He was just about to help a ewe that was in difficulty lambing when out went all the lights. He was accompanied by his 4-year-old son, William. The two of them ran back to the house to fetch a torch and were soon back inside the shed ready to lamb the ewe.

The new gates have completed and modernised the sheep building.

'Hold the torch steady son, hold the torch steady,' said Harry, and he soon brought out a live lamb. 'Hold the torch steady son,' he said, and again Harry put his hand into the ewe and brought out another live lamb. The little boy was amazed. 'Hold the torch steady son,' Harry repeated and pulled out a third lamb and then a fourth. William was now totally amazed and stood with his mouth wide open. 'Do you think it's the light that's attracting them, Dad?,' he asked.

Chapter 8

Insurance

When I bought my first field, I was in no rush to get it insured. It was a mass of weeds with no buildings. No one is going to steal a field. There is only a small chance of someone setting fire to the hedge or crashing a car into the hedge or fence, and so I didn't get any insurance for some time, but looking back I probably should have done. However, when work was started by contractors to get the field into shape and build the poultry house, I realised I had to get insurance.

If you look on the internet, there are numerous companies that provide insurance for farmers – it is not just the National Farmers Union (NFU) that will insure you. I opted for the NFU, as they have got a good reputation of paying out claims. I have always thought that many insurance companies are keen and eager to receive your money to pay your premiums, but when it comes to paying out, they may drag their feet and try to get out of it; so in my opinion, whatever insurance company you choose, you must read the small print on the policies. Be careful: in some insurance policies, the big print giveth and the small print taketh away! The first task is to telephone the insurance company and then get one of their representatives out to see your property – they should be well experienced and able to advise you on what to insure and the most economical way of going about it. However, don't forget it is their job to sell insurance, and you don't want a representative who is just getting as much from you as possible, so be sensible, think things through and get other quotes from other insurance companies – after all when you insure your car, you would probably not jump in and accept the first quote.

I insured my poultry house, bulk feed bin and auger, tools and machinery, and sheep building against fire, theft, storm and flood, explosion, earthquake, aircraft collision, escape of water, impact, riot and malicious persons. My growing hedge and gate were insured for fire, aircraft explosion and theft. The hens were insured against fire, theft, explosion, earthquake, riot, malicious persons and electrocution. I also insured produce and deadstock, which would include my eggs, hay and straw, seeds, chemicals, fuels and fertilisers against fire, theft, explosion and aircraft. Feed on the premises is also insured plus eggs in transit. Money is insured which may be kept on the premises, plus money in transit and then, heaven forbid, malicious attack – which includes death, loss of limbs, permanent total disability, temporary total disability, temporary partial disability and damage to personal effects. Check each section of your policy for any excess you have to pay on your claims. It doesn't end there because I have also got employer's liability, public and product liability, and environmental liability.

A neighbour of mine sadly had a farm fire. She telephoned the fire brigade: 'Come quickly; it's Grange Farm, which is quite isolated.' 'How do we get there?,' asked the lady on the switchboard. 'Don't you have those red things with the sirens on any more?,' asked my neighbour in her panic. The fire was found to have been started deliberately. Claire, the farmer's wife, telephoned the insurance company on the day of the fire. 'Our barn was insured for £30,000, and so we want to be paid out immediately,' she reported. 'It's not quite that easy,' replied the man from the insurance company. 'We first assess the value of your barn and provide you with a new one of comparable worth.' 'Well if that's the way you do things, you can cancel the policy on my husband immediately!,' she replied very sharply.

Public-liability insurance is needed if you get a public-liability claim. This is when a member of the public has suffered injury or loss as a result of your negligence. This could include any sheep getting out on to the road and causing an accident. Public liability is not a legal requirement, but I would advise you to insure because an incident could be very costly indeed. A usual figure would be insurance for £5 million in any one incident. In this environment that we now live in, if a burglar gets hurt while breaking into your property, would he sue you?

I am also insured for product liability. This means if you sell some goods, and they are faulty, your customer could make a claim; for example, if I sold some eggs that were proved to give food poisoning, the customer could make a claim.

Once you start employing someone for full time, part time or even voluntary work, you will need employer's liability insurance. This is a legal requirement. This insurance protects you against any action taken against you where an employee has suffered injury or loss as a result of your negligence. The limit of indemnity on this insurance is usually £10 million. You will be issued with a certificate of employer's liability, which must be displayed at your place of business. I am also insured for environmental liability, which could include fly tipping on my fields or perhaps someone putting chemicals into the watercourse.

Livestock can be insured for many things, which can get expensive. My livestock are insured for fire, aircraft, explosion, electrocution, earthquake, riot, malicious persons, theft, and salmonella in the laying flock. You can also insure for mortality, straying, transit, worrying by dogs and even fertility cover and specialist diseases such as foot and mouth, TB and brucellosis.

When the first outbreak of foot and mouth occurred on 20 February 2001, I immediately telephoned my insurance company to insure against it but because an outbreak had occurred, they refused to insure. On the arable farm, growing crops are not usually insured against the weather – but they can be insured, for example, if the combine catches fire and some of the crop gets burnt; and the harvested crops can also be insured in the buildings for fire, theft, flood and other options – because the farmer would lose income on crops he could not sell.

If you put all your insurance on one combined policy, it will probably be cheaper than lots of different policies. Personal accident cover is very useful in that if you did have an accident, you would be paid, and also you could be paid out for loss of income. An insurance man tried to sell an accident policy to another one of my neighbours, who shall remain nameless because he is not very bright. The insurance man was not getting very far, when he asked his final question: 'How would your wife carry on if you died?' 'Well it would be no concern of mine as long as she behaves herself while I am alive' was my neighbour's reply. The farmer did not take up the offer of this policy!

Agricultural motor vehicles and trailers should be kept in good order and are covered by the Health and Safety at Work Act. Most large and expensive machines will need insurance. Tractors can be insured for the following:

1. Third party – this is limited to legal liability to third parties such as injury to the other road user.
2. Third party fire and theft – this covers any third party, and you could also make a claim if your tractor was stolen or damaged by fire.
3. Comprehensive insurance – this gives the most cover. It insures for damage to your own vehicle, damage to the other person's vehicle, fire and theft, and legal liability.

Most new tractors would be insured fully comprehensively, especially if they are held on lease. Some really old models may be third party, but comprehensive is obviously the one to go for. If your tractor is registered for use on the road, it must be insured for at least third-party road risk or SORN'd. If you never use it on a public highway, it must be SORN'd. If you use your tractor just to cross the road only a few times a year, you still must be insured under the Road Traffic Act.

On the cattle market car park, you will see lots of farm vehicles pulling livestock trailers, not many Range Rovers, plenty of Land Rovers and plenty of four-wheel-drive trucks. In my opinion, some of these trucks are not worth insuring (they have to be of course) because no one would ever steal them for a number of reasons:

1. They won't travel very far because there would be very little diesel in the fuel tank.
2. The owner is the only man in the world who can fiddle with the door handle to enter the vehicle – it certainly wouldn't be locked, and the driver's window would be held up with gaffer tape.
3. It would be very difficult to drive fast because (a) it will only achieve 30 miles an hour and (b) you run the risk of all the equipment falling out of the back such as fencing posts, bale string, rolls of wire, lengths of chain and rope, a drum of diesel and a large round bale.
4. The cab will want mucking out, owing to the copious amounts of mud, muck and straw on the floor, and even some grass growing next to the accelerator pedal and almost enough soil to grow a crop of potatoes, plus the hole in the floor will be a health hazard because of the exhaust fumes. It would be littered with receipts, illegible paperwork, cans of marker spray, a box of castration rings, and various spanners and tools.
5. The bent bumpers, broken tail lights, broken wing mirrors and one door being a different colour from the rest must all deter thieves.
6. The safety belts will be the only good thing because the owner will not have used them.

A typical outside view of a farm vehicle.

Interior of vehicle.

When you build or buy your house, you will want the house plus the contents insured. The building probably wants insuring for:

1. Fire, smoke, explosions, earthquakes and lightning.
2. Theft.
3. Aircraft or anything dropped from above.
4. Storm and flood.
5. Collision in case a vehicle crashes into it.
6. Malicious people or vandals.
7. Labour and political disturbance, civil commotion and riot.
8. Falling branches or trees.
9. Falling television aerials or falling closed-circuit cameras.
10. Water or oil leaks.

Additional insurance will be available, and people's needs are of course different. Contents of the house including money probably need to be covered as above except for labour and political disturbance, civil commotion and riot. Additional cover could include contents in your garden, office furnishings and equipment which are used for your business, and payment for the cost of replacing documents and security certificates if they are damaged in your house; also, fatal injury to you, your wife, husband or partner if you are fatally injured inside the territorial limits as a result of:

1. A fire in the house.
2. An accident in the home or garden.
3. An accident while travelling on public transport.
4. An assault in the street.

There are many different things you can insure. I was pleased the contents of our freezer were insured one hot summer day because the freezer had broken down, and we were left with a mess. Our claim was settled in full, which was the cost of replacing the food. You can even be insured if you are emotionally stressed, owing to damage to your property such as from fire.

Public liability is also advisable on your house insurance as well as personal and legal expenses. If you are with the same insurance company for a number of years, your cover may be eligible for a no claims discount. Also, if you agree to stay with the same insurance company for a number of years, you should get a discount; this is called a long-term undertaking.

We've got the strength of the insurance company around us.

Chapter 9

Rules and Regulations for Keeping Livestock

Please note that rules and regulations in the UK do change from time to time, so treat this chapter as a guide only – you must check with DEFRA for up-to-date information.

Unless you are going into growing plants – either market gardening or specialist horticulture or you have a large farm that is suitable for arable farming – then you must have livestock. You will want livestock if your heart is in the job, but once you start keeping them your life will change forever owing to the time and commitment that you need. Keeping livestock is a 365-days-of-the-year job, and when you first start your farm, you may not be able to employ help. Feeding and looking after stock are much easier when you are younger, and as you get older it becomes harder to use a muck fork. A fellow that came to buy sheep off me to start his first flock asked me what qualifications I had. I replied by telling him I had an MSc 'What – a Master of Science?,' he said. 'No,' I replied, 'a muck-shovelling certificate!'

Livestock have always been a large part of my life – I grew up with them. I have always been committed to looking after stock. As a boy, I didn't want posters of pop stars or footballers; I just liked animal pictures. I would look after my animals before and after school, at weekends and during the holidays, and didn't want to go on long holidays because of having to leave them. I would go and stay with my uncle and aunt on the farm, and they taught me well – to check stock and to look out for ailments. My uncle told me what to look for when checking feet, ears, eyes and backsides. Of course, when you have got livestock, you will also have some dead stock. Twelve months full time on a farm plus working with my uncle and aunt and my pets at home gave me enough confidence to keep livestock when I began teaching Rural Studies at Higham Lane School, Nuneaton. I taught for 25 years, keeping a flock of pedigree sheep breeding lambs, British Alpine goats and Jersey cows, rearing pigs for half pigs for freezers, rearing calves until they were 6 months old, plus rabbits, guinea pigs, chickens, ducks and geese. As I have got older, I have shown pedigree sheep, commercial sheep, finished lambs, goats, Jersey cattle, eggs and even a bale of hay, and as a boy I showed rabbits, guinea pigs and bantams.

When you start farming, you could keep beef cattle, calves, sheep, goats, alpacas, pigs or poultry. Keeping a dairy herd is unlikely unless you have a lot of land or buildings, but you could keep a house cow or two – that is, a cow that produces milk just for your family or to feed calves.

You will need various forms and paperwork, which are legal requirements before you start keeping your livestock. This is what you will need: first, you will need a County Parish Holding (CPH) number. Once you have got this, you can then apply for everything else, but before you do I would suggest contacting your local vets, and make sure they deal with large farm animals. Get registered with them, because from day one with your stock, you may need to purchase veterinary medicines, and of course you may need a vet in a hurry if one of your animals has an accident or if you need assistance lambing a ewe.

When you move any farm animal, you will need an animal-movement record book, which can be obtained from DEFRA. In this book, you record movements off the holding showing the date, the species, the number of animals moved, individual identification number for each animal, CPH location the animals were moved to, haulier's name, registration number of the transport vehicle and a section for your own use. This book should be filled in within 36 hours of the movement taking place. Movement records must be kept for 6 years and can now be held electronically or in written form. An inspector from Animal Health has been out to check my records at least twice over the years and found nothing wrong. You will also need to fill out a medicine book for all of your livestock. In this, you record the date of purchase of the medicine, the name of the medicine and quantity purchased, the batch number, the identity of the animal group treated, the date treatment started, the date treatment was finished, total quantity of medicine used, length of withdrawal period (number of days), earliest date for sale of animals or produce (milk, meat or eggs) and who administered the medicine. The medicine book can be obtained from agricultural supplies shops such as Countrywide.

Paperwork for Cattle

You will need a CPH number before you can get started. You will need to register your herd with DEFRA, and you will then be issued with a herd mark and registration document. All cattle now have to be double-tagged and have a tag in each ear. You must have a primary tag, which is the large yellow tag, and a secondary tag, which could be the same as your primary tag or a metal tag or button tag. The information on the tag is as follows: the crown logo, UK, six-digit herd number followed by a number, which is a 1, 2, 3, 4, 5, 6 or 7, followed by the animal's individual herd number. The large tag has space for more information such as an individual management number. With beef calves, you have 20 days to double-tag your animals, and with dairy calves, you have to double-tag within 20 days of birth; the secondary tag, however, must be placed in the calf's ear before the calf is 36 hours old.

You don't need a licence to move cattle, but you do need cattle passports. Cattle movements are controlled by the British Cattle Movement Service (BCMS). These passports are enforced so that the whole life of the cattle can be traced. Each passport gives details of the animal, records of where it has been throughout its life, and details of its death. To apply for a passport, an application must be received before a calf is 27 days old. Each movement must be reported either online or by telephone to BCMS.

Passport applications must be made within 7 days of tagging, which is a maximum of 27 days in total from the date of birth. A cattle passport remains with the animal throughout its life, and it can't enter the food chain at the abattoir without one. If a disease such as foot and mouth breaks out, these records could be very valuable indeed, helping to stop the spread of disease.

Once you have moved your cattle, you must not forget to fill in your movement book, and if treated for ailments or disease, all treatments must be recorded in your medicine book. Once you move cattle on to your farm, you are on a 6-day standstill – that means you have to wait 6 full days before you can move animals off your property, unless they go straight to slaughter or the farm has isolation facilities.

Bovine Tuberculosis (TB)

This is an infectious disease of cattle that can be spread from cow to cow by the bacterium *Mycobacterium bovis*. This disease is causing much heartache, stress and anxiety for cattle farmers, and eradication of this disease is only a long-term aim. The aim at the moment is to stop the disease spreading and getting worse. This bacterium can also infect and cause TB in badgers, deer, goats, llamas, alpacas, pigs, dogs, cats and many other mammals. Badger culling in parts of Britain is aimed at controlling the spread of the disease.

The way cattle are tested is by carrying out the tuberculosis skin test. The test is carried out in the neck by a veterinary surgeon or a trained technician. The ear number is recorded, the hair is clipped, and the thickness of the skin is measured. Then, the skin is pricked in two places. The top site is injected with avian tuberculosis and the bottom site with bovine tuberculosis. Seventy-two hours later, the veterinary surgeon or trained technician returns, and the site is checked for lumps. The lumps are measured using callipers, and any changes are recorded. The animal will be either negative or positive to the test, or maybe the result will be inconclusive. It is a wonderful feeling if all the animals are clear of the disease, but disappointing if 200 cattle or so are tested, and the very last one reacts positive to the test. An inconclusive result means that the animal is retested in 60 days' time – you can't move the animal until you get a clear result.

A positive result means that the animal has to be sent for slaughter, and the farmer is compensated, but a special animal such as a prize winner or a good 'stock getting' bull is a great loss and could have a long-term effect on the herd. The herd is then placed under movement restrictions until every beast in the herd has passed two other tests 60 days apart. This means that although the farmer may want or need to sell cattle, he can't, so he will use more food, fodder and labour keeping the animals on the farm. If he is selling store cattle or in-calf heifers, they may be getting too old, and the in-calf heifers may even calve before the movement restrictions can be lifted. Negative cattle that are on the same farm as cattle with positive results may be taken directly to an abattoir for slaughter or possibly to restricted farms that can fatten negative cattle, but for pedigree animals worth a lot of money it is a great shame just to get slaughter prices. Cattle with positive results can't be sold.

If each animal passes the second test, the movement restrictions are lifted, but the herd is further tested 6 and 12 months later. Hopefully, the herd will remain clear of the disease, and testing will continue to be routine for that area.

High-risk areas are tested annually. These would include farms in the South West, West and East Sussex, but much of England is now tested every 2 or 4 years. Please note: rules and regulations do and have changed, so contact DEFRA to get up-to-date information on the disease and find out from them how often you will be tested, and certainly check the number of days that you will be under movement restrictions.

Cattle can be expensive to buy – my old farm boss, Mo, was short of money. 'John – I don't know if I should buy a bicycle or a new Friesian cow,' he pondered. 'You would look silly riding the cow,' I replied. 'I would look even sillier trying to milk the bicycle' was his witty reply.

Paperwork for Sheep

An executive agency of DEFRA is the Animal and Plant Health Agency (APHA). You will need to contact them for a flock number. This number will be unique to your flock and shows where the sheep were born. All sheep must be tagged, and your ear tags will have this number on it plus consecutive numbers starting at 0001 and then going on and on into the hundreds or even thousands. Different breed societies require different information on the tags. The Kerry Hill sheep society has the letters KHS, your individual flock number, the consecutive number for the individual sheep and the year the lamb was born, e.g. 2013.

Nowadays, sheep must be tagged within 6 months of birth if housed overnight, or 9 months of birth if not housed overnight, or whenever they move off the farm of their birth if this is sooner.

If you are to keep your sheep for longer than one year, you have to put in two ear tags, both have the same individual unique number. One of these tags must be electronic. However, if your sheep are going for slaughter before 12 months of age, you can attach just one year tag showing only your flock or herd mark. This tag must now be electronic.

Sheep born between January 2008 and January 2010 will have two identical tags – one in each ear or a tag in one ear and a tattoo in the other.

Sheep born before January 2008 will have just one ear tag showing the flock number, but if they are no longer on the farm where they were born, they may have a secondary tag called an S Tag.

Sarah tagging a Kerry Hill lamb.

All sheep keepers must have a flock register – the modern ones incorporate movement records. In this book, you record replacement ear tags, date of identification of animals, and births, deaths and an annual inventory. Once a year, you will have to record the number of sheep that are on your holding.

To move sheep to an abattoir or market or on to another farm, or to bring animals on to your farm, you will need a movement licence. This was called form AML1 and was obtained from Trading Standards. The new Animal Reporting and Movement Service (ARAMS) is provided by SouthWestern from their Milton Keynes office and has now taken over responsibility for sheep/goat/deer movements from local authorities in England in 2014. The system allows you to record your movements electronically or using the hard copy form. To record your movements online, you must register for a new account with ARAMS, which costs nothing. I must admit I am using the hard copy form. Both systems record the CPH number of the place of departure, the keeper's name, the full postal address of the holding of departure, the individual identification or flock mark, the departure date, the time the first animal was loaded, the time of departure, the expected duration of the journey, the transport details and details of the receiving location. The licence is for traceability so that you know the exact movements of the sheep, and then if disease breaks out, all movements can be traced. Once you move your sheep onto your farm, you are on a 6-day standstill, which means you have to wait 6 whole days before you can move animals off your property (unless you have isolation facilities or they go straight to slaughter).

I did suggest at a Kerry Hill council meeting that perhaps we could cross our Kerry Hill sheep with elephants and kangaroos, my idea being that the babies born would have large enough ears to put the tags in and a pouch to carry all the paperwork.

Paperwork for Pigs

Paperwork applies to all livestock, pigs included, so you can't get started as a pig keeper until you receive your CPH number. You need to register as a pig keeper with DEFRA; the Animal and Plant Health Agency (APHA) will do this for you. You will be given a herd mark, which consists of one or two letters followed by four digits.

Nowadays, you have to prenotify all pig movements; pigs were moved on an AML2 paper form, but this is now obsolete. Prenotification for moving pigs can be done electronically. This service is operated by the British Pig Executive (BPEX) and is known as the eAML2.

For pig keepers without access to a computer, pigs can be moved by notification via telephone, fax or post through the Meat and Livestock Commercial Service Ltd (MLCSL). This is a paper bureau service, and they will put your information onto the electronic system for you.

You have to register your details such as name, holding number, etc. on the BPEX website or telephone the MLCSL; they will give you details on how to register, and once registered you can start to move pigs. You prenotify BPEX or MLCSL of the move. You need to print off a copy of the haulier summary sheet, and a copy must accompany the pigs during transport. The new keeper of the pigs at the destination holding must confirm details of the movement within 3 days of the move taking place either electronically direct to the eAML2 system or by writing to, telephoning or faxing the MLCSL. The haulier retains the summary sheet that he collected on loading. If pigs are to be moved to a market or collection centre, you still need to prenotify the movements, if the market or collection centre agrees to notify BPEX on the day of the pigs' arrival. I would advise you to check with BPEX or your local authority before moving your pigs for more details or any changes that have taken place. As with other livestock, you will need to keep your animal movement record book. Once a year, you will need to record the number of pigs that are on your holding. Records will have to be made available for inspection.

Once you have your new arrivals, you will be on a standstill. If you bring pigs on to your premises, you can't move pigs off for 20 days unless you have isolation facilities or unless they are going straight for slaughter, but you only have to wait 6 days before you can move cattle, sheep or goats off. If you buy or take in cattle, sheep or goats, your pigs will need to be put on a 6-day standstill.

Pigs are identified by ear tags, tattoos or slap marks.

1. Ear tags start with the letters UK followed by your herd mark.
2. Tattoos – the letters UK are not required, just your herd mark.
3. Slap marks – this is a permanent ink mark that is put on to each shoulder of the pig; again, the letters UK are not needed, just your herd mark.

A pig slapper is a head attached to a long handle. The head consists of tattoo needles, which penetrate the skin with tattoo ink. You wield the slapper at the pig and literally slap the pig with it. Many old-fashioned slappers were poorly designed, and a number of pig keepers have slapped their own legs with a lovely tattoo!

Pigs that are under 12 months old can be moved with a temporary paint mark, e.g. a red dot or a blue line if you just move them between holdings, but for slaughter or market they must have an ear tag, tattoo or double slap mark.

Pigs over 12 months old can't be moved with a temporary paint mark and must have permanent identification. You can't remove or replace identification numbers without permission from DEFRA unless you can't read it or it is for welfare reasons. If they are pedigree pigs, you have to notify the British Pig Association (BPA).

All this record keeping and form filling, as for all other livestock, is done in case of disease. If disease breaks out, DEFRA can see exactly where the pigs originated from, when they have been moved and where to – and of course the difference between bird flu and swine flu is that for bird flu you need tweetment, and for swine flu you need oinkment!

My old farm boss, Mo, used to tell the tale of moving a pig before 'all these fancy regulations,' as he put it, came into force. He had a sow that lived in the orchard at the front of the farmhouse. He had just got married but no honeymoon or fancy hotel for him and his new bride; it was a quick wedding, a small reception and then back to the farm. He did say that they could lie in bed as a treat, as he hadn't had a lie in for 15 years. However, they could only lie in bed if his sow was just lying down and not rolling in the mud. If she was rolling in the mud she would be on heat and would need taking to the farm next door in the wheelbarrow. So, they went to bed and the next morning, Mo told his new wife to get out of bed and look through the window. 'Is she rolling in the mud or just lying down?,' he asked. 'Rolling in the mud' was her reply. Mo got up and dressed, and put the sow in the wheelbarrow to take her next door to Bill's Large White boar. The sow was served, and Mo returned with her. He told his wife they would get their lie in bed tomorrow. Tomorrow came, and again Mo sent his wife to look through the bedroom window. 'Is she rolling in the mud or just lying down?,' he asked. 'Rolling in the mud' was her reply. 'I'll take her next door again – we won't get our lie in, but it is good stockmanship to have her served twice. Again he got up, got dressed and put the sow in the wheelbarrow. He took her to Bill's boar, and again the sow was served. On the third morning Mo said,'I am confident we can have our lie in bed this morning. Go on girl, get out of bed and have a look out of the window,' he ordered. 'She should be lying down – is she?' he asked. His wife was silent. 'Well is she rolling in the mud?' At last his new wife replied. 'No, she's not lying down and she's not rolling in the mud.' 'Well, what *is* she doing?' roared Mo. 'Sitting in the wheelbarrow' was her reply.

Footnote – a sow rolling in the mud is not necessarily on heat.

Paperwork for Poultry – Layers

If your flock of hens numbers 50 or more, you need to register them with DEFRA, even if you have them for just part of the year. DEFRA encourages poultry keepers to register voluntarily; if registered, you can be contacted if there is a serious outbreak of disease. You can register by telephoning the DEFRA helpline, where an advisor will fill in the form for you over the telephone. You will be sent the completed form to check, or of course you can register online through the DEFRA website.

All poultry entering or leaving your premises must be recorded in an animal-movement book; again, if a disease outbreak occurs, the poultry can be traced. Once registered, you will get a producer number, and all eggs sold retail and at public markets within the EU must be stamped with a code. This code identifies the method of production, e.g. organic, free range, barn or caged, as well as the county of origin and the laying establishment. The number starting with 0 UK identifies eggs from organic production produced in the UK; 1 UK – free-range eggs produced in the UK, 2 UK – barn eggs produced in the UK, 3 UK – enriched colony system eggs produced in the UK.

If the eggs are both organic and free range, then the stamp must be 0 UK and not 01 UK. After this part of the stamp comes a five-digit number. This is the producer identification code, so an example on an egg would be 1UK13491 (my number). Eggs may be stamped at the production site or at a registered egg-packing station. If the eggs are not stamped at the production site, the distinguishing number as well as the producer's name and address should be shown on all containers of ungraded eggs leaving the production site and on all documents.

You do not have to stamp your eggs if you sell them directly to the consumer for their own use, if eggs are sold to customers coming to your farm or door to door, or possibly at a local public market.

If you have fewer than 50 birds and sell at a market, you do not have to stamp your eggs, but you must put up a sign showing your name, address, best-before date and advice on how to keep them chilled after they have been purchased. However, some markets insist on all eggs being stamped.

Stamped eggs can be traced, which is useful for the egg inspectors to see where they have gone in case of an outbreak of disease. Egg inspectors are responsible for enforcing and monitoring legislation. I sell my eggs to two main packers, and so the eggs leave here with no stamp on them. When they are graded at the packing stations, my number is stamped onto each egg.

If you have more than 40,000 poultry, you must apply to the Environment Agency for a bespoke Environment Permit. This agency regulates intensive poultry farms. They monitor and control pollutants, and look at removal and treatment of waste matter.

There is now a National Control Programme (NCP) for salmonella in laying flocks. Salmonella infection in eggs can cause food poisoning. This programme is followed except when eggs are produced for private domestic consumption only or if the holding has fewer than 350 hens and supplies direct to the consumer or via local retailers.

All large flocks are sampled in the following ways:

1. Chicks. Boot-swab the chicken house a week before the chicks arrive. A swab consists of a piece of gauze that is moistened, opened out and used to wipe an area where the birds live. Both sides of the swab are soiled. You attach the swab to your boot or shoe and walk in the area. Remove the boot swabs carefully, place in a sealed bag and send off to an approved laboratory.

2. Rearing of pullets. Two pairs of boot swabs for floor rearing. For cage-reared birds, one large composite faeces sample is taken, which should be fresh.

3. During lay. Two pairs of boot swabs for barn or free range, first between 22 and 26 weeks of age and then every 15 weeks during the production of eggs.

As well as doing the tests yourself, the NCP for salmonella requires an official sample to be taken annually from one flock on all holdings with more than 1000 birds. This will be done by an egg inspector. This test may replace the routine 15-week sample taken by you or farm staff, and the control sample would also require the collection of dust. If you pack your eggs on the farm and sell to shops and restaurants, you will also need a packer's number from DEFRA.

Paperwork for Broilers

You need to register your flock, under the Great Britain Poultry Register at DEFRA, and all movements of birds must be recorded. DEFRA has published the *Guide to the National Control Programme for Salmonella in Broiler Flocks*, available through their website. The NCP in broiler flocks applies to all producers, but there are two exceptions that apply to many farmers.

1. Producers who have fewer than 2000 birds present at any one time, and where the producer sells direct to the consumer, e.g. sale at the farm gate or selling to local retailers who supply house holders.

2. Birds are sold for private domestic use – these birds are not sold on the market.

Salmonella testing is as follows. The operator/producer samples all flocks, and the tests are done within 3 weeks before the start of depopulation. This is done with two pairs of boot swabs (or hand-drag swabs in houses with fewer than 100 birds). Also, there is an official sample on 10% of holdings with more than 5000 birds – which is two pairs of boot swabs or hand-drag swabs from one flock of broilers.

There is some flexibility in testing on farms with good salmonella control. You can apply for derogation not to sample all of the flocks on your farm. Check with your local Animal and Plant Health Agency to see if you meet all of the criteria.

Samples are sent to the laboratory on the day they are taken, but in exceptional circumstances they can be sent later – which must be within 48 hours of being taken. During this extra time, samples are to be kept at 4°C.

If your results come back positive, and salmonella is detected, contact your veterinary surgeon and DEFRA. Poultry meat from a flock that has been tested positive can be sold for human consumption, which I find very surprising. The producers must tell the processor of the positive result, and the processor must then control the hazard and minimise the risk of other meat being contaminated. If you intend to kill your broilers on the premises and prepare them as oven-ready, you will need to comply with animal welfare, meat hygiene and all the statutory regulations. You will need to check with DEFRA for all the up-to-date rules and regulations.

Paperwork for Turkeys

You will register your flock with DEFRA, and all movements of birds will be recorded in your movement book. DEFRA have published the *UK Guide to the National Control Programme (NCP) for Salmonella in Turkey Flocks*, available from their website. It lists statutory requirements for the monitoring and control of salmonella. To get the full picture, you need to read it, but for a fattening flock of turkeys, the main rules are as follows. The NCP regulations only apply to fattening flocks with 500 turkeys or more and breeding farms with over 250 birds. On these farms, birds are tested by the producer 3 weeks before slaughter.

Samples are sent to an approved laboratory usually within 24 hours of conducting the test, but it must be within 48 hours. All compulsory test results will have to be kept for at least 2 years and be made available for inspection. If salmonella is detected, contact your vet and DEFRA. DEFRA produce an explanatory leaflet on poultry meat regulations, entitled *Poultry Meat Special Marketing Terms*. This should be read before starting, to keep your turkeys for meat, but many of these regulations do not apply to the small producer.

Turkey farms with an annual throughput of 500–10,000 birds that are sold locally are exempt, and so these farms will not be required to carry out their own operator sampling for salmonella. However, these farms may be selected at random to have an official sample – 10% of holdings are selected for this. DEFRA also recommend voluntary sampling.

To carry out the test, there must be either two pairs of boot swabs or one pair of boot swabs and 900 cm^2 of dust swab samples. These may be pooled at the laboratory for testing, or in flocks of fewer than 100 birds, four hand-held 900 cm^2 dust swabs are sufficient if boot swabs are impractical. If you intend to kill your turkeys on the premises and prepare them as oven-ready, you will need to comply with animal welfare, meat hygiene and all the statutory regulations. You will need to check with DEFRA for all the up-to-date rules and regulations.

Transporter Authorisation

If you travel over 65 kilometres – approximately 40 miles – in connection with your farm business or commercial activity that aims at financial gain, you need to hold transporter authorisation. There are two types:

1. Type 1 is for journeys over 65 kilometres and up to 8 hours' duration.
2. Type 2 is for journeys over 65 kilometres and more than 8 hours' duration.

These authorisation certificates are issued by the APHA and last for 5 years unless they are revoked or suspended. You and any of your staff who will transport livestock need to be trained. You need to take an assessment test, which is a 'tick box' test, before you will be issued with a certificate of competence. To find out the venue for the test and dates and times, etc., you can contact your local APHA. Application forms are available online from the government's website.

If you transport long distance (Type 2), your vehicle has to be approved – again, contact your local APHA to find out where to get the vehicle approved. These long distance trips must be supported with contingency plans, which deal with any emergencies such as breakdowns or animals becoming ill, and you also need a tracking system, which records details of the journey set out in a log.

Rules and regulations are obviously in place so that livestock are transported as humanely as possible with journeys not causing unnecessary pain or suffering. There are different rules for transporting cattle, sheep and pigs, and there are a lot of them, so check with DEFRA to make sure you abide by them. The general common sense rules are as follows:

1. Plan your journey to be as short as possible.
2. Stock should be fit enough to travel and be treated so they get as little stressed as possible.
3. The vehicle for transportation should be safe and designed for the purpose of transporting livestock with enough room.
4. Ventilation is very important – heat and humidity can cause considerable stress.
5. On long journeys, rest must be made available, and also water and food if appropriate.
6. The vehicle or trailer must be thoroughly cleansed and disinfected within 24 hours after it has been unloaded.

Our son, Jonathan, aged 8, transporting a hen with no relevant paperwork.

Chapter 10

Making a Start with Cattle

Susie had an iron cow
She milked it with a spanner
The milk comes out in shilling tins
The little ones a tanner!

This rhyme was around before decimalisation of money; a shilling was 12 old pence, and a tanner was 6 old pence.

To make a start with cattle, keeping a few calves is easy and practical, but I will cover calf rearing, taking your calves to store weight or buying in stores, beef production, dairy farming and keeping a house cow. However, remember that cattle do produce a great deal of muck, so if you just rent a small shed in the corner of a field, and you have no land of your own, you need to find a farmer willing to take it, and he probably will not pay for it, so you will have to pay to have it moved.

Calf Rearing

Calves can be reared by either:

1. Artificial feeding.
2. Calves being allowed to suckle their mother or foster mother.

Artificial Rearing

The first and most important factor when rearing calves is to make sure your calf has drunk colostrum, at least 3 litres in the first 6 hours of life. The antibodies in the colostrum protect the calf from many diseases. It is also rich in energy and protein, and it contains vitamins A and D. The very first milk after calving contains the highest levels of antibodies. It tastes good, too: I have eaten many a rice pudding and egg custard made with colostrum instead of milk, but these days it would be frowned upon.

The calf should receive colostrum for the first 3 or 4 days of life. However, it should not receive too much colostrum, as it can cause scours (diarrhoea). On some farms, alternative colostrum dried from cows' natural colostrum is purchased, as Johne's disease can be spread by the mother's colostrum.

Calves should be at least one week old before purchasing them. You can buy them at this stage or older – perhaps 12 weeks – or buy older store cattle.

A top tip to get on the farming ladder is to buy or rent a suitable building on a farm, buy in a few calves at 1 week old, rear them and then sell them at 12 weeks old. If you only start with a few, do your homework – visit markets to see what 1-week-old calves are being sold and for what price, and see what older calves are making. Work out how much food they will consume, and if the market looks right, have a go. If you make some money, you can use this to buy more calves; this will obviously be done part time, so you can look after your calves before and after work, and you will need very little capital.

You can buy privately direct from a farm, which is good because you are not mixing batches of calves, or you can buy from a calf group such as Livestock Link Ltd at Leamington Spa, Warwickshire. Firms such as this take some of the bother out of buying and selling. They are a cattle marketing supply chain that charge commission on transactions. They will deliver even batches of young calves direct from local farmers and will supply the details of the calves' sire if you are part of a certified beef scheme. These schemes make you meet certain standards, and then you will get extra money for the calves. Calf groups such as Livestock Link Ltd will also supply reared calves from local rearers, which you can inspect on the farm before purchase. They will also buy your finished cattle, often the ones they supplied. These will go into high-class butchers or to large supermarket customers.

Death losses are of course the responsibility of the rearer after a company guarantee period – calves are usually just guaranteed for the first week. If you sell to a group, they will expect the following:

1. All the relevant paperwork to go with the calf.
2. A well-grown calf for its age, e.g. a target weight of 140 kg at 16 weeks.
3. The calves should all be weaned.
4. Bull calves should be disbudded and castrated if you are asked to do so.

5. The coat should have a good bloom.
6. The calf should be bright-eyed and not scouring.
7. There should be no signs of disease or parasites.

Economically, it is better to rear large groups of calves because haulage charges for a single calf will eat into your profits, as will buying just the odd bag of calf feed – bulk buying is much cheaper.

Calves waiting to be sold at Melton Mowbray market.

You need to choose calves that are breathing normally and showing no signs of scouring; look also for a shiny coat and bright eyes. The navels should also be healed. You can buy pure Holstein bull calves – these are a by-product of the dairy industry and may well be out of pedigree cows. I prefer buying something more 'beefy' and more expensive – a good continental cross that will go on and be a first class beef animal would be my choice, e.g. Simmental, Limousin and Charollais crosses. Calf prices might be cheaper from September to May when there are more calves for sale than during June to August.

Get organised before your calves arrive; the calf house must be light and well ventilated. It should be cleaned out well and disinfected, and have a good, clean bed of straw. I think the disinfectant is very important indeed, especially before your next batch of calves. Before your calves arrive, make sure you have purchased your feed, buckets, feed troughs and hay racks. On arrival, give them 'Rispoval' – an intranasal vaccination against viral pneumonia (they are then best injected against viral pneumonia at 12 weeks old).

Calves can be housed in individual calf pens often up until weaning, which prevents them from sucking one another. They can be fed milk or milk substitute out of a bucket. After weaning, they can be put into groups. Group them according to size, weight and sex. I wouldn't suggest more than 10 to a pen straight after weaning. Plastic polymer calf hutches have become popular – these are tough hutches that look similar to large dog kennels and are placed outside – you put wire or hurdles at the front so that the calf can spend time in or out. You can get individual hutches or hutches for a group of calves. Manufacturers claim that they almost stop scours and pneumonia, as they are draught-free and well ventilated with low humidity.

Plastic calf hutch.

Instead of bucket feeding, you can use plastic feeders. These are similar to troughs and consist of compartments into which you pour the milk (one compartment for each calf). The calves suck from the teats attached to the feeder. The feeder hangs on the calf pen side or gate. Three-teat and five-teat feeders are popular; once one group are fed, you can move the feeder to the next pen. One manufacturer claims that by using a feeder with five teats, you can feed over 70 calves in an hour.

Calves drinking from a plastic feeder.

Milk or milk substitute is usually fed at approximately 10% of the calf's birth weight, e.g. 4 kg of milk per day for a 40 kg calf, which is best in two feeds. The way I feed calves in a practical way is to start them on an average of 1½ litres per feed and increase this to 2 litres per feed twice a day.

Milk or milk substitute can be fed warm (body temperature) or cool at 10–25°C. Opinions differ on this, but I prefer feeding warm milk. Calves on restricted feeding will consume about 15–20 kg of milk substitute.

Ad lib feeding is when calves can just go and help themselves to feed. You may use 30 kg of milk substitute on this system, but the calves will probably have put on more weight. I prefer restricted feeding because you can see what each individual calf has drunk. In this modern high-tech world in which we live, some large farms now have computerised calf feeders. These machines recognise each individual calf. They know its age, what time of the day it can feed and how much milk it is allowed. If you have just started farming, you can manage without one.

Calves are usually weaned at 5–6 weeks old. I prefer to go the extra week and wean at 6 weeks. Hay or straw and concentrate feed should be introduced from the first week of life; 18% protein starter pellets are probably the best, and at weaning they will be eating 1.00–1.25 kg per head per day. Post-weaning, feed a 16% or 18% coarse mix; a good mix may include rolled barley, flaked maize, linseed, palm kernel, beet pulp and minerals. If you grow your own cereals, your ration can be home-mixed, which will save you money.

Disbudding a calf. Cattle are better with their horns removed; this avoids damage to other livestock and also people. Horn buds are cauterised using a dehorning hot iron, or alternatively a dehorning paste can be used.

Weaned calves.

After 12 weeks, calves will be eating more hay or silage. Calves born in the autumn will spend the winter indoors and then go out to grass in the spring. Calves born in the spring often stay in the building during the summer, autumn and winter, and are turned out the following spring.

It is important to disinfect your building very thoroughly once your calves are sold. Disease can build up in the building, and so resting the building between batches of calves is also good practice but is not always practical. First-time farmers have lost money resulting from disease in the calves, so you must be as hygienic as possible.

Single Suckling

This is when the cow calves and is left to rear her own calf; the calf may be left on the cow for 7–10 months. These are called suckler herds. Cattle will calve in the spring or autumn. After weaning, the calf will be sold as a store or taken on to finishing as a beef animal.

Our friends, Philip and his son Nigel, make money by keeping suckler cattle. They keep them as follows. Calves are born in March or April (and some in the autumn), and these spring-bred calves grow on well to make excellent beef. They are born inside in covered yards; their mothers would include Blonde-d'Aquitaine cross Limousin, and their sire would be a Simmental, which makes them a very strong beefy animal. They would receive 1 kg of feed per day, a home-grown mix of wheat, barley and oats mixed with minerals and silage. The calves would be creep-fed on a home mix. Into the

Single suckling.

mixer goes 700 kg of corn (oats, wheat and barley mixed), 100 kg of soya, 100 kg of ground beans and 100 kg of biscuit meal (waste from a biscuit factory). Mothers and calves would stay in the yards for at least 3 weeks after calving. Before turning them out, they would be disbudded and the bull calves castrated. After turn-out, the cows would still receive 1kg of feed per day and the calves still creep-fed. The cattle will be wormed at least once during the summer. The calves will be weaned during October and November, and will come into the cattle yards at more or less the same time. The calves would be wormed again and the cows kept separate.

The cows will continue to be fed and will be calving again during March and April; then, young stock will not go out on grass again but will be kept in the covered yards. During later life, they will eat up to 5 kg of feed and silage per day and will be ready for slaughter at 22 months old – the heifers would weigh 500–550 kg live weight and the steers 600–650 kg live weight. You can copy this system with just one or two cows or a few more, but your costings will be more if you have to buy in all your concentrate feed. The above example will make more money with home-grown cereals.

Double- or Multi-Suckling

This is when the cow can rear her own calf, plus you buy in additional calves for her to rear. Cows can be brought inside to suckle the calves twice a day, or some rear their own calf plus another one out in the field. A very good mother can rear many calves in one lactation – perhaps four or even eight.

Store Cattle

Once your calves are reared, they can be sold as reared calves or kept on as store cattle and sold as beef. Store cattle are cattle that will be sold to another farmer to finish them as beef or you can buy store cattle and take them up to beef weight. If you are buying store cattle, they need to be as inexpensive as possible, but at the same time be as good as you can afford. You could buy cattle, which may be dairy cross steers or heifers at various ages and weights. You then need to be looking at selling them for beef at round about 500–700 kg live weight.

A group of store cattle that match do look the part if you are selling or buying. They want to be the same size and weight, and the same sex; they should look as if they have got the potential to grow and be healthy. A poorer, smaller animal in the batch will reduce the value of the batch – the smaller ones should be marketed together.

This is the type of store animal to buy – good strong animals that will produce good quality beef.

A very good example of how to rear store cattle is the way my friends Clive and Geoffrey do it, and so do many other farmers. They make money doing it this way and I would recommend it. First, – buy animals about 300 kg, 9–12 months old, during October. These should be beefy calves that have the potential to be good beef animals with no traces of Holstein or other dairy breed qualities.

These bought cattle will have come from a suckler herd and not from a dairy herd. They are likely to have been sired by a Simmental or Limousin, and their dam would also be a beef type such as a Lincoln Red or Beef Shorthorn, which may also be crossed with a continental breed. Most animals this age will have a bolus inside the rumen, which releases wormer all summer. If not, you will be advised to put one in in the spring.

You turn them out to grass when you get them home and give them just a maintenance ration of 0.680–0.907 kg (1½–2 lb) of quality beef nuts a day. You keep them outdoors for as long as you can, but if it comes a lot of wet weather, they will churn the grass up into mud, and they will look thoroughly miserable.

Once in a covered yard, carry on with 0.680–0.907 kg (1½–2 lb) of beef nuts plus silage, and if you can buy some potatoes, carrots or parsnips cheaply, then do so. This is easier if you live in an area like Lincolnshire. Give them plenty of straw bedding and water.

Turn them back out on to grass in April when there should be some grass growth. Drench them with a combined wormer and fluke drench or give them a worming bolus, and carry on with 0.680–0.907 kg (1½–2 lb) of beef nuts per day. By October, the heifers will no longer be store cattle, but most will kill as finished beef, which you would sell then at market or to a butcher. They should weigh about 500–600 kg live weight and be about 2 years old or just under.

The steers now want selling on – still as stores but big stores. These will then be purchased by a farmer, who will keep them in covered yards. They would be fed ad lib with rolled or milled barley, beef nuts, silage and possibly potatoes, carrots or parsnips. This farmer would keep them for about 6 months, and then they would be ready for the butcher and weigh about 600–700 kg live weight and be about 29 months old. Cattle are best finished for beef before they are 30 months old. Cattle over 30 months have to go to specialist abattoirs where the spinal cord and vertebrae are removed, which is routinely done now ever since the BSE and CJD outbreaks.

Big, strong store cattle ready to be sold on to be finished as beef.

An alternative, slightly more intensive beef system is the 18-month beef-production cycle. Calves are born in the summer and autumn, and kept in covered yards during the winter; they go out to grass for just one summer, are brought into the yards in autumn and stay in the yards until they are finished beef animals, then killed in the spring at 16–20 months old after eating ad lib concentrates and silage. A sign found in a Herefordshire farm read, 'It won't cost you any money to cross my land but my bull might charge!'

Beef carcasses.

Intensive Beef Production

I have some more friends, Tim, Colin and Andrew, who are very good at producing bull beef. They keep a breed of dairy cow that originated in Holland and Germany called the Meuse Rhine Issel, which gives less milk than a Holstein, but they produce a sturdy beef calf with good confirmation and good feed conversion, especially when crossed with a beef breed such as the Simmental. As well as my friends, many other bull beef producers farm bull beef in the following way. Calves are weaned at 5 weeks old, by which time they will be eating plenty of coarse mix and straw (not hay). At 8 weeks old, their ration would consist of half-milled barley and coarse mix fed ad lib, and at 12 weeks old milled barley and beef nuts mixed with chopped straw (4 parts barley to 1 part beef nuts, which works out at 14–15% protein). They will grow at approximately 1.4 kg a day.

Finished bull beef.

The bulls are kept indoors all the time and at 14 months would be sold for beef at about 700 kg live weight. During this time, one bull would eat approximately 2 tonnes of barley and concentrates. If not castrated, bulls grow faster and convert feed into muscle quicker than castrated cattle. As they are indoors, they do not get wormed. The building must be substantial enough to withstand the strength of entire bulls. Bulls can be more dangerous than castrated animals, so you need to be aware that they could turn nasty. They also must be kept separate from heifers. Two things that would help you make money on this system are (1) if you are able to breed your own calves and (2) growing your own barley.

Horse meat has been found in burgers and pies, and has been sold as beef. This scandal, of course, made national headlines. However, one leading supermarket representative was quoted as saying: 'Our meat is of very high quality and has had to clear a number of hurdles before it goes on sale!'

Beef Cattle Breeds

There are many breeds of beef cattle – their respective breed societies will give you the information you need to know. I have selected a few.

Traditional British Breeds

Aberdeen Angus

Originating in Scotland, this beef breed is famous and sold in Waitrose, Marks & Spencer and Burger King. Customers are willing to pay a premium for the excellent quality meat, which is marbled and very tender.

Aberdeen Angus bull.

The breed is all black and naturally polled, and calves easily. It has a quiet temperament and will finish on grass. Very good beef is produced when crossed with other breeds. Aberdeen Angus bulls are used to cross-breed, especially with dairy cattle to cut down on difficult calvings.

Belted Galloway

These cattle originate in Galloway on the west side of southern Scotland. Crossing the Galloway with the Dutch Belted cow, the Lakenvelder, in the 17th and 18th centuries is believed to be the origin of the Belted Galloway, but no one is certain. The colour

Belted Galloways showing the three different colours.

of the coat is usually black, but dun and red animals are also found. They have a white belt, which runs right around the middle of the body.

It is a hardy breed and can live outside all winter, even in tough weather conditions. They are slower to grow than some breeds, finishing at 24–30 months old on forage alone. They are polled and live a long time, with some cows producing calves at 14–15 years old.

Hereford

Hereford is a wonderful-looking British breed. The body is a lovely deep red colour with a white head and crest, white dewlap and throat, white underneath and a white brush to the tail. They can finish on grass, and the meat is marbled, which helps to provide flavour.

Hereford bull.

When crossed with other breeds, it will always sire a calf with a white head. When an American visitor from Texas visited my school and looked over the cattle-shed door where we kept two Hereford calves, he said to my students, 'I see you keep Hereford cattle in England too, do you?' One of my 12-year-old students was very quick to give him the history of the breed.

Other native beef breeds include the Galloway, Highland, Lincoln Red, Red Poll, South Devon, Sussex and Welsh Black.

Continental Breeds

Charollais

Charollais originated in France and was imported into this country for the first time in the late 1950s. It is creamy white in colour but can be a wheaten colour, and it is a large breed. It is an excellent terminal sire, especially when put onto good suckler cattle. Crossbred cattle are

Charollais bull.

quick to grow, and they have a deep chest and strong muscular hind quarters. They can finish at 1 year old and give excellent beef when crossed with many breeds. Several butchers favour this breed.

Limousin

Limousin originated in the west of the Massif Central between central and south-west France. This is a large hardy breed, first imported to Great Britain in 1971 and of a golden red colour. They are easy to get in calf, producing good milk for their calves, and they calve easily. This breed is quick-growing and has an excellent carcass, producing good-quality beef with a high killing out percentage. It is crossed with many different breeds giving excellent beef.

Limousin bull.

Simmental

This breed originates from Switzerland. In some countries, it is a dual-purpose breed, but in Great Britain, it is used for beef. The colour varies from gold to red with white. The head is white, and there is often a white band across the shoulders. They are docile, easy to handle and are hardy. They calve easily, make good mothers and milk well. The pedigree cows and bulls live a long time, and they are found all over the world with a population of 41 million. It is crossed with many breeds giving very good beef.

Simmental bull.

Other continental breeds include the Belgian Blue and the Blonde d'Aquitaine.

Dairy Farming

Cattle are kept for meat and milk, and some of this milk will be turned into butter or cheese. It is hard these days for a first-generation farmer to start at the very bottom of the farming ladder with a large dairy herd. In the 1930s and 1940s, a viable dairy herd consisted of 20 or 30 or so cows, but these days you really need at least 80–100 cows. When I left school in 1970, I worked full time on a farm for a year, and during this time, I milked around 20 cows in a cow shed. At that time, it was classed as a very small herd. Things moved on, and later when I was a college student, I worked part time milking about 120 cows in a modern herring-bone parlour.

To set up a dairy farm today even on a rented farm takes up a great deal of capital. You need enough land to keep the cows, and you need to buy the cows and equipment associated with them. However, you would probably get contractors to do most of the tractor work such as silage making and manure spreading. This type of farming is only possible for the first-time farmer who has capital – it was certainly easier to do this many years ago. Finding a county council farm to rent is probably the best way to start dairy farming.

Herring bone parlour.

I went to school with Ian; he was a few years younger than me, but we used to walk to and from school together each day. His parents rented a 20.23-hectare (50-acre) Warwickshire County Council farm, milking cows in an old-fashioned cow shed, rearing heifers and breeding pigs. They were first-generation farmers.

At weekends and during holidays, we used to play with our 'Britains' farm tractors in the pig meal bin. The family took on another council holding bordering the existing unit and then another one that also adjoined.

These days, Ian is married to Karen, and they have a grown-up son, James. All three work very hard indeed on the farm; they are absolute enthusiasts in dairying, keeping 140 Holstein cows with good buildings and a modern milking parlour. They strive for the maximum milk yield possible, rearing their own heifers and carrying out a thoroughly good job.

The calves are taken away from their mothers at birth; most don't drink from the cow (it seems to be a fact with their Holsteins that they don't find the milk). These days, on many farms, the calves are not allowed to drink from the cow because of the transfer of Johne's disease from the mother to the calf; the mothers' colostrum is fed for the first 3 or 4 days of life from a bucket, and the calves are then given cows' milk, not substitute milk. The heifer calves are housed individually and bucket-fed on warm milk twice a day. The bull calves are housed in groups and fed from plastic feeders with teats.

The bull calves are sold at 5 weeks old - not castrated or disbudded - to a calf group, Livestock Link in Leamington Spa, Warwickshire, and I have been told by David from Livestock Link that they do an excellent job of calf rearing. Calves are encouraged to eat concentrates (ad lib; 18% protein pellets) as soon as possible and wheat straw (not hay or barley straw).

Heifer calves are weaned, disbudded and grouped into about eight per pen. Sometimes an odd one may get an infected umbilical cord (even though it was treated with iodine at birth), but this is soon cured with antibiotics. Scours and pneumonia would be encountered occasionally.

These heifer calves soon grow, and at about 5–6 months old they will be moved into larger groups. They are not turned out to grass until they are at least a year old. They will now be eating about 3 kg of concentrates each per day and will start to be fed silage.

Once turned out to grass, they receive no concentrates. The heifers continue to grow well, and just before bulling they are vaccinated for bovine viral diarrhoea, two injections

3–4 weeks apart, and a magnet is placed in the rumen to catch any wires that might get swallowed.

The in-calf heifers are wintered indoors and given a blend of feed including wheat, soya, sugar beet, brewer's grains plus silage. When they are big enough, they are served with DIY artificial insemination (AI; carried out by the stockman, not an AI man) which is when they are about 15–17 months old and so they start calving at 24 months old.

Some have embryo transplants placed in 7 days after bulling by a technician who comes to the farm. A pregnancy diagnosis is carried out on every cow at over 7 weeks into pregnancy. About 30% would need help at calving; the dairy cows are housed in cubicles, fed once a day with a total mixed ration (a blend of feed and silage formulated for maximum milk production) and fed in the buildings and not in the parlour. They are milked three times a day in a herring-bone parlour at 7 am, 3.30 pm and 11 pm, and don't go out into the fields to graze.

The target is for each cow to produce over 50 tonnes of milk per cow before she is sold. Heifers yield about 10,000 litres per lactation, and the herd average would be 11,500 litres per lactation. The cows calve every 18 months and not every 12 months (or thereabouts).

Cows are put through a foot bath of formalin three times a week. Foot problems do occur, including digital dermatitis, but foot baths certainly help. Good cattle are sold at specialist auctions including a few bulls. The enterprise is a credit to them all.

House Cows

I was a school teacher for 25 years, and during this time I kept numerous animals on the school farm. I purchased a pedigree Jersey calf, called Windsor Coronets Crystal 6th, aged just 2 weeks old, from HM The Queen. She was to be our house cow. We handled her a great deal, halter-trained her and showed her at various agricultural shows winning prizes with her. After she had calved for the first time, she was no trouble to milk with a machine because she had been handled so much. She went on to have heifer calves, and we also kept Higham Lane Natalie as a house cow.

Heifer calves that are intended for the dairy herd are not pushed with masses of feed. I made a mistake by keeping Crystal on substitute for too long, and she was too fat, but she slimmed down by the time she was ready to be artificially inseminated. A few years later, we purchased another Jersey heifer from HM The Queen, called Windsor Grand Good News, who grew up to be a tremendous house cow.

Heifer Windsor Coronets Crystal 6th, which I purchased from HM The Queen on behalf of Higham Lane School.

One of my students at school asked me, 'What kind of milk comes from a forgetful cow?' 'I've got no idea,' I replied. 'Milk of amnesia, Sir!' was his reply. The whole class groaned!

On average, you need about 0.404 hectares (1 acre) of land to keep your house cow. This could be large enough to keep her calf and last year's calf, but if your land is very poor, you may need more land than this. On modern dairy farms, often where the cows are milked three times a day, the cows don't go out but remain in covered yards all year round. I do like to see the cows grazing in the fields, and they do like being out in good weather. When cows are turned out in spring for the first time after a long winter, it is a delight to see them full of excitement running around the field.

You can keep your dairy cow out all the year round, but come early winter, in my opinion, she will be better in a covered yard giving your grass a chance to rest. A covered yard is best, but you will need a separate section where she can be tied up and milked. This section needs a concrete floor for easy cleaning.
I made the right decision in starting out with a calf and getting her used to being handled. Cows are herd animals and like company, so Crystal lived with some beef calves. I wouldn't go to a market and buy a Holstein that has been used to being milked in a parlour. Temperament is the key factor because you don't want to be fighting a bad-tempered cow that kicks you and kicks your bucket of milk over. However, if she does kick your bucket of milk over, don't cry over spilled milk, turn the udder cheek, and moove on!

The two breeds that I would recommend are the Jersey and the Dexter. If you have only got one or two cows, I wouldn't go for the highest milk yield possible by feeding them with masses of cattle cake. Enough milk for the house and to feed her calf is easy to achieve by feeding plenty of grass with some concentrates and then in the winter – hay and silage with some more concentrates. Many modern dairy herds go for a high-input high-output system – keeping Holsteins with a lot of concentrate feed and milking three times a day. If you have enough acreage, you can grow some of your own foods, especially hay and silage, and perhaps some barley.

When your heifer is mature enough to be served you will need to purchase some semen and get her artificially inseminated. It is no use keeping a bull for just one or two cows. Get organised and purchase your semen from a reputable company such as Genus UK. They will store the purchased semen and send out the AI man to inseminate your cow. At the Warwickshire Farmers' dinner and dance, various agricultural awards were given out, and I am pleased to say that our local AI man was awarded the 'long service' award!

When your heifer or cow comes on heat, it is termed 'bulling.' Signs of bulling are as follows: she will stand to be mated by other cows, she will be restless, she may jump on the front end of another cow and she may have a mucous discharge with blood from the vulva called a bulling string. If blood appears from the vulva, it may mean she was bulling perhaps 48 hours ago, and so you are too late to have her served this time. She will also bellow when bulling, especially if living alone. I have found that if I stand and touch my toes in front of her (taking care and looking carefully behind me) she may try to jump on my back. I have done this on many occasions but have not heard of others doing it (I am not surprised!) and am not recommending that anybody else tries it! I am sure the Health and Safety brigade would not approve of this probably quite dangerous practice, so please do not attempt it.

If you think you have detected bulling, make a note of it in your diary and start looking for it again in about 18 days' time. She will be bulling for about 12 hours. If you have only one heifer or cow, watch out for the bellowing. Once she is artificially inseminated, you need to check to see if she is bulling again in 18 days' time. If she is, she will need to be artificially inseminated again.

A friend bought a house cow, and he borrowed a bull from the next-door farm. The cow was bulling, but she kept moving away from the bull. The bull tried time and time again to mount her, but she moved every time. If he approached from the front, she would move, and if he approached from the back, she would move – it made no difference, and she would not tolerate him. My friend asked the vet for his opinion. 'Did you buy the cow from Nuneaton, Warwickshire?' asked the vet. 'Yes,' replied the farmer, 'How on earth did you know that? You really are a brilliant vet!' 'No,' replied the vet 'it's just that my wife is from Nuneaton!'

Calving

A cow is in calf for 280–290 days. When she is due to calve, her udder will become much larger – which we call 'bagging up.' She will become restless, will want to go off on her own and will stop eating. You will notice a clear fluid coming from the vulva, the pin bones open and ligaments around the tail head relax. The water bag comes out, bursts and releases fluids. The heifer or cow will strain to push the calf out. Normal presentation of a calf is the feet first, followed by the head and chest. Once past this stage, the calf should slip out easily.

Problems – How to Help a Cow Give Birth

Some calves will take some getting, and lubrication will help considerably. Lubricant can be purchased from your veterinary surgeon or farm supplier such as Countrywide. If she is not progressing about an hour after the water bag appears, I would investigate and have a feel to see what is going on. Make sure your hand and arm are clean, as you don't want to infect your heifer or cow.

A large calf may need a calving rope attached, or you can use a specially manufactured calving aid. I wouldn't start pulling at the calf until I had found two front feet and the head or two back feet and the tail. You pull when the cow strains. I would pull in a slightly upward or horizontal fashion until the head and chest are out and then pull downwards to get the rest of the calf out. It certainly takes the hard work out of calving using a mechanical calving aid.

As you calve more cows, you become more experienced. You will get a few problems such as a calf with its head back, one leg back, breach or twins. If the calf isn't breathing, clear mucous from the nose and mouth, poke straw up the nose, move the front leg rapidly, massage the chest or give artificial respiration by breathing up its nose. The cow should get up and lick the calf – if not, get her up and encourage her. Afterbirth is usually expelled soon after giving birth.

After our son, Jonathan, was born I got in the lift and pressed the button for the ground floor. I immediately recognised an ex-student in the lift; it was Simon. 'Hello Sir – are you a new dad?,' he said. 'Yes,' I replied. 'Our first child, a boy – Jonathan. How about you?,' I enquired. 'Yes. Our first child, a girl, but it took ages and ages for her to be born. We had two midwives and a doctor, but it still took hours and hours – I'm exhausted! I said to the team – what are you messing about at? If my old Rural Studies teacher was here, Mr Terry would have got some lubricant and some calving ropes, and the job would have been done in about 20 minutes.' 'You didn't really say that, did you?,' I asked. 'Yes,' he replied, 'but my wife didn't like me saying it!.' 'Well I wouldn't have said it in front of my wife,' I replied.

Your cow wants to be served 2 or 3 months after calving. Don't forget that she is in calf for approximately 9 months, and you want her to produce one calf per year. You will need to tie up your cow to milk her twice a day. A chain around her neck will secure her. She will need a manger or trough in front of her for feed and an automatic drinking bowl. Her faeces will need to be shovelled into a wheelbarrow, and after each milking you need to wash out with a hose pipe.

The above brings back happy memories of my year working on a farm after leaving school. I worked full time on a mixed farm. I generally milked cows on my own – only about 17 or 20 of them, and they all had names – Friesians and Ayrshires tied up in a straight line in the cowshed. On hot days, we would throw buckets of water on the concrete floor to cool it down. The cows were milked by machine, and the bucket, which was part of the machine, would be tipped into a milk churn. My farm boss, Mo, would joke and say 'After you've taken the churns outside, leave the tops off if it is raining!'

You will need a portable milking machine. I have milked by hand – I used to milk the school goats by hand, which is easy, but a cow does make your wrists ache. Old farmers do say you get used to it; in fact, all cows were milked this way before machines. I certainly wouldn't want the hassle of hand milking. Your machine needs to be kept clean. When I milked the Jersey cow on the school farm, it took almost as long to clean the machine as it did to milk the cow. An old farmer was milking his Jersey cow, and he noticed a large fly fly straight into the cow's ear. The fly then appeared as if by magic in the milking bucket. Did it go in one ear and out the udder?

A major disadvantage of a house cow is if you want to go on holiday. You may be lucky enough to find someone to feed a few calves or sheep, but to milk a cow twice a day for a week needs someone willing and experienced. It is easier to let her calf suckle her milk from her. You can do both, but beware if you milk your cow and then leave some milk for the calf to suckle – she will be wise to you, and she will try her best to hold back the milk for her calf. If you want to do both, then milk the cow out completely, take what you want for the house, then feed the rest to your calf. If you still have some left over, buy another calf. Milk straight from the cow is rarely drunk these days – most milk is of course pasteurised, i.e. heated to a minimum temperature of 72°C for 15 seconds. It is possible to sell raw milk, but there are strict rules, and it can only come from TB-free herds. Cheese made from unpasteurised milk must be labelled 'Made with unpasteurised milk' or 'Made from raw milk.' Raw milk cannot be sold in Scotland, but unpasteurised dairy products can be sold.

Your dairy cow needs to be dried off about 56 days before she calves again. At this time, she will be well in calf and will need this rest period before her next lactation. If the dry period is less than 40 days, she will give less milk in her next lactation. Milking is stopped, perhaps for the evening milking, and you can then milk her again the next morning. After missing a few milkings (there is no strict rule), take her away from the milking shed, and keep her in a yard with just straw to eat and water. She will need a dry cow therapy tube inserted into the teat canal in each quarter.

Dairy Cattle Breeds

There are a few breeds of dairy cattle – their breed societies will give you the information you need. I have selected a few below.

Ayrshire

This is a lovely-looking dairy animal originating from Ayrshire in the south-west of Scotland. They can be any shade of red or brown with white, and it doesn't matter which colour predominates.

Ayrshire cow.

They do not produce as much milk as Holsteins, but the top herds will average over 8500 litres per lactation. They are easy to manage, live a long time and are good, healthy cattle suffering from few diseases. They will do well under poor conditions. This breed is a favourite of mine after the Jersey. I always think a field of red and white Ayrshires looks terrific. They convert grass into milk very well and are found in most parts of the world.

British Friesian and Holstein

Holstein cow.

The British Friesian originated in Europe and became popular in the 19th century in Great Britain. These are smaller than Holsteins and carry more flesh. They produce less milk than a Holstein but produce bull calves that can be slaughtered as good beef. Holsteins are a strain of the Friesians imported originally from North America.

Pure Holsteins are the highest-yielding dairy cattle in the world, and the breed is now the predominant dairy breed in the UK. Both British Friesians and Holsteins are usually black and white with some red and white ones.

Jersey

Ideal as a house cow, Jerseys are my ultimate favourite breed of cow, and I love visiting agricultural shows to see the best examples of the breed. Jerseys, as the name suggests, are originally from Jersey in the Channel Islands. They are usually fawn in colour but can be all shades of brown or grey or almost black. They sometimes have white patches, and they usually have a black tip to their tail, but occasionally they can have a white tip. They weigh about 400–500 kg. They can produce about 5000 litres of rich milk per lactation at about 5.8% butterfat and 3.8% protein. I have made some excellent butter and cheese from Jersey milk. Jerseys can be crossed with beef breeds to produce acceptable beef. Many pure-bred Jersey bull calves are slaughtered at birth because they don't make good beef cattle. Some bull calves, however, are slaughtered at about 5 months old for rose veal.

Jersey heifer Windsor Grand Good News purchased from HM The Queen. The Queen is seen here visiting my students and myself at school, the highlight of my teaching career.

Other dairy breeds include the Meuse Rhine Issel, Guernsey and Shorthorn.

Dual-Purpose Breeds

Dexter

Ideal as a house cow, Dexter cattle originated in south-west Ireland and are descended from black cattle kept by the Celts. They are the smallest breed of cattle in Great Britain – about half the size of the Hereford breed. A cow will only measure 92–107 cm at the shoulder and weigh 300–350 kg. They are black, red or dun in colour. There are two types – short-legged and non-short-legged; they can be polled or horned.

It is a dual-purpose breed, which means it will milk well and be good for beef. An advantage of the breed is that you can keep more cows on a small area of land than larger breeds. Milk yield would be about 2500–3000 litres in a lactation, with some cows achieving over 4000 litres. The milk is rich, and the beef is excellent, popular in farmers' markets and shops.

Dexter cows and calves.

Heifers are quick to mature and can be put to the bull at 15 months old, calving at 2 years old. Dexter cattle are friendly, hardy and ideal for the farmer with a small acreage.

Chapter 11

Cattle Diseases and Ailments

If you suspect signs of a notifiable disease you must immediately contact your local Animal and Plant Health Agency.

Acetonaemia or Ketosis

This is a metabolic disorder. It is found in high-yielding dairy cows on an inadequate diet and occurs during the first 3–6 weeks after calving. High-yielding cows have high energy demands, and so with this complaint their demands are greater than the energy intake. So, if they are not taking in enough energy to compensate for the energy lost in the milk, we say the cows are in a negative energy balance. The liver functions are upset, resulting in circulation in the blood of ketones, (ketones are a class of organic compounds produced from rapid fat breakdown). Animals will lose their appetite, refuse to eat concentrates and lose weight, and milk yield drops. There is a characteristic smell of ketosis in the breath, which is a sweet, sickly, pear-drop smell. A small proportion of animals may froth at the mouth or lick objects such as cubicle barriers.

To cure this condition, a glucose replacement is necessary either by intravenous injection or by oral administration. To prevent it, feed the animals well, but don't get them over-fat. Cows that are over-fat at calving or that have other problems such as milk fever, mastitis or hypomagnesaemia are at increased likelihood of getting acetonaemia.

You should never change the diet suddenly; introduce change gradually. After any change in diet, it takes up to 2 weeks to stabilise the rumen.

Actinobacillosis or Wooden Tongue

Bacteria get into the mouth through small wounds, which may have been caused by eating rough, sharp food. Cattle of any age can become infected, but it is more common in cattle over 1 year old. The animal will be in pain, the tongue becomes very thick, hard and swollen, and the animal will have difficulty swallowing and chewing the cud. The tongue may protrude from the mouth, and so food may be dropped out of the mouth. The condition can be treated with antibiotics if you catch it in the early stages, but if it is left to the later stages, the cattle will probably die.

Actinomycosis or Lumpy Jaw

Bacteria get into the lining of the mouth through small wounds, which may have been caused by eating rough, coarse straw or other feedstuff. The bacteria is fungus-like, it causes swellings which are hard and are found on the jaw bones with a swelling under the jaw. The swelling can rupture and discharge yellow smelly pus. Animals will be in pain, salivate more than normal, cannot eat properly and have a fever. Treatment is possible with antibiotics, but often the animals are culled, as many never recover properly.

Anthrax

Anthrax is a notifiable disease caused by bacteria and can occur in man. The spores are very resistant indeed. Thankfully, it is not a common disease in Britain. Ingestion of contaminated feed or soil will spread the disease.

The first sign of the disease is probably that the animal is found dead. However, you may see them swaying on their legs, and after death dark blood may bleed from the mouth and nose, and sometimes the anus and vulva.

Animals must be destroyed and buried on the farm, along with bedding and contaminated soil, and any other material that is contaminated.

Bloat

Bloat is quite common in cattle. This occurs when gas collects in the rumen and reticulum, but cannot escape, thus making the rumen swell. The upper-left side of the abdomen will be distended with fewer ruminal contractions. If the distension progresses so that the upper-right side of the abdomen is also distended, the animal will show abdominal pain, bellowing, staggering, collapse and possibly death.

Frothy Bloat

The fermenting food in the rumen forms a foam, and the gas it contains cannot be expelled. This can happen when the cow has been eating rich grass in spring or autumn, particularly if it has a high clover content or when eating lots of rich concentrates. The animals can die in a short time, and the condition can affect a high number of animals.

Mild cases may be treated with cooking oil, but it is best to consult your vet. Obviously if you get an outbreak of frothy bloat, you get the cows off the grass as soon as possible and give them straw or hay to eat.

Free Gas Bloat

Gas is trapped in the rumen because something is obstructing the oesophagus such as a potato. A gastric tube may be used to push the object down, or a trocar can be used to puncture a hole in the rumen – this is not done very often, as it can introduce infection.

Another cause of bloat could be lack of ruminal contractions, and the nerve supply to the rumen may be damaged owing to digestive problems, or even swallowing a foreign object, such as wire.

High-risk pastures are best grazed when the weather is dry and if the animals are to be turned out on to lush pasture, they should be given straw or hay before hand, restricting their grazing time and gradually increasing it each day. Some individual cattle are more prone to this than others.

Bovine Viral Diarrhoea (BVD)

BVD is caused by a pestivirus. Affected cattle suffer reproductive problems, abortion, infertility and suppression of the immune system. If the cow is infected during the first 3 months of gestation, the foetus will also be infected because its immune system is not fully developed and it often dies at this stage. If born alive, we say that the calf will be persistently infected (PI) with the BVD virus and is immunotolerant to the virus. Infected animals will pass the virus onto other cattle by direct contact but may appear normal. Some, however, may develop mucosal disease – a variant of BVD that is usually fatal.

This disease is the biggest problem in cows in early pregnancy, often causing death to the foetus and thus abortion, or weak calves born prematurely. Sometimes infected bulls can also spread infection. When BVD virus is spread among young stock, it suppresses their immune system, making them susceptible to pneumonia and scours. Dairy-farming friends in this area vaccinate against the disease, but PI animals should be culled.

Brucellosis Contagious Abortion

This is a notifiable disease but has not occurred in the UK for some years. Bacteria are responsible and infect the womb and placenta. The calf will be born prematurely or born weak and often dies. Calves that live may spread infection. The foetal membranes will be retained, and aborted calves will spread the infection, as will the afterbirth, discharges and milk. Laboratory tests confirm the disease. All infected cattle must be slaughtered. The disease can infect humans, causing undulant fever.

Bovine Spongiform Encephalopathy (BSE)

BSE is also known as 'mad cow disease.' It is a chronic disease that affects the central nervous system. It was first diagnosed in Great Britain in 1986 and is a notifiable disease.

The agent that causes the disease is not a bacteria or a virus: it is an abnormal version of a protein found on cell surfaces called a prion protein. The prion changes and then destroys the brain and spinal cord – the brain actually becomes spongy. It is not contagious and so does not spread between cattle. The disease appears when the animal is 4–5 years old, and it affects cows; beef cattle are slaughtered for meat at a much earlier age and so it does not affect them.

The first symptoms are when the animal acts in a strange way; it may stumble and hold its head awkwardly and stand alone. As the disease progresses, the animal may become aggressive, may have tremors under the skin and will lick its nose and become disorientated, and of course milk yield will have dropped.

Animals become infected by eating animal feed containing these prions; this food may contain the remains of cattle with infected brains and spinal cords. There is no cure for the disease. It is now illegal to feed potentially infected feed to farm animals including prohibited processed animal proteins such as mammalian meat and bone meal. Check the APHA website to get up-to-date information. Catering waste can also not be fed including waste from factory canteens and restaurants.

Animals that have the disease have to be slaughtered – compensation is available. There are also various rules surrounding offspring of BSE-infected cattle and cohort animals of BSE cases – check the DEFRA website for up-to-date information.

If a person eats beef from an infected cow, the person is at risk of getting a variant called Creutzfeldt–Jakob disease (CJD) which is a very rare disease affecting the brain. At the time of writing, BSE has almost been eliminated from cattle in the UK, but vigilance at this stage is very important.

Two cows are talking in a field. One cow says to the other 'What are your views on this "mad cow" disease?' The other one replies 'It's no use asking me, I'm a duck!'

Choke

This is when the animal has a solid object such as a potato or apple blocking the oesophagus or throat. The animal cannot swallow, and it may stretch its head forwards and have its mouth open; if it cannot release gas, this can lead to bloat. Other symptoms include coughing, drooling and perhaps some feed around the nostrils. The animal could die from bloat or an infection could set in where the object is obstructing.

Digestive juices and saliva may dissolve enough of it to be swallowed, but you are more likely to have to remove it. You can try to move the object up the oesophagus from the outside, then gag the animal and put your hand down the throat and pull the object out. This can be very dangerous without correct restraint and equipment owing to the animal's very sharp and powerful cheek teeth. A gastric tube can be pushed into the oesophagus to try to push the object into the rumen (a sedative may help in relaxing the oesophagus muscles). If these methods fail, a trocar can be pushed into the abdominal and rumen wall to relieve the gas.

Clostridial Diseases – Caused by Bacteria

Blackleg, bacillary haemoglobinuria, pulpy kidney, malignant oedema or gas gangrene, and botulism are clostridial diseases. Clostridial diseases (except for botulism) can be vaccinated against. Blackleg can be vaccinated against on its own, or a combined vaccine can be used, which covers a number of clostridial diseases. I will look at two of the above, Blackleg and bacillary haemoglobinuria, in more detail below.

Blackleg

This is a disease affecting cattle 6–18 months old. The bacteria are found in the soil, so animals exposed to soil, such as earthworks, ploughed land or a soil floor in a building could be vulnerable. The bacteria are ingested by the cattle, and then live in the muscles, usually in the animals' legs, remaining dormant until the animal has an injury, such as from a fight, causing a reduction in oxygenation of the muscle cells. The bacteria increase in numbers very quickly indeed, producing a gas, and toxins enter the bloodstream. Infected animals are often found dead. After the animal has died, the muscles become darker in colour (hence the name Blackleg), and they smell awful. The disease can be treated if caught very early.

Bacillary Haemoglobinuria

The first sign is probably finding the animal dead. The liver is damaged by migrating liver fluke, probably during late summer and early autumn. There is no treatment. You can try to cure animals with antibiotics, but the animals will probably die.

Fatty-Liver Syndrome

This occurs mainly in the first 2 weeks after calving, but it can be later than this. As the name suggests, it affects overfat animals. High-yielding dairy cows will have become fat during late pregnancy because of overfeeding. They lose too much body condition too quickly and lose appetite at the same time, which aggravates the situation. A negative energy balance results, and the liver cannot convert fats into glucose quickly enough – fat then accumulates in the liver and reduces its functions, which brings on hypoglycaemia. Lack of glucose means the animal will produce ketones, which leads to ketosis. Clinical signs may include other problems such as milk fever, mastitis and ketosis, plus rapid weight loss, reduced milk yield and nervous symptoms. The animal could have retained foetal membranes, and finally death could occur owing to liver failure.

Treat with a bolus of intravenous glucose plus oral propylene glycol. Prevent it by not having the cow too fat before calving. Avoid stress and group cows together that are at the same stage before calving, so that their feeding requirements can be met.

Foot-and-Mouth Disease

This is a notifiable disease caused by viruses, affecting cattle, sheep, deer, pigs and goats. Infection can be carried by many means including humans, hay, straw, birds and vehicle tyres.

Animals have a fever, milk yield drops, they drool saliva and become lame. Ulcers are found in their mouths, between the claws of the feet and on the teats. All affected animals must be slaughtered, plus contact animals slaughtered then buried or burnt. This is the most heart-breaking disease of all.

Freemartinism

When twins are born where one is a heifer and one is a bull, the heifer may not be able to be bred from. The blood supply in the mother is shared by the twins. The male hormone responsible for the development of the male reproductive system is first to become active and gets into the heifer calf.

This means it will have an underdeveloped and small female reproductive system and a small vulva, and, when older, will not come on heat; 90% of heifer twins will be affected. These heifer calves usually go for slaughter.

Infectious Bovine Rhinotracheitis (IBR)

This is a contagious disease that affects the respiratory and reproductive organs. It is caused by a virus 'bovine herpes virus 1' (BHV1). Animals infect each other by fluids from the nose, vagina or semen. Some cattle can live with the virus for years without showing symptoms. There are two forms: respiratory and reproductive.

Respiratory Form

This is inflammation of the upper respiratory tract. Animals may show discharge from the nose, conjunctivitis, ulcers in the mouth, fast shallow breathing, coughing, high temperature, loss of appetite and death in some cases. Outbreaks of IBR are often a precursor to more serious outbreaks of pneumonia.

Reproductive Form

There will be a loss of milk yield, a discharge of pus from the vulva and reddening of the vaginal lining and the animal will urinate more often. Some animals may abort, and calves may be born weak. Bulls may have an inflamed penis.

Most animals recover, but they can still carry the virus, and so they should be removed from the herd. Antibiotics won't treat the virus, but if a bacterial infection is present secondary to the virus, antibiotics can be used to treat the bacterial infection. Vaccination is available.

Johne's Disease

This is caused by a mycobacterium, and there is no treatment or cure. Calves are infected during the first few weeks of life, but the disease doesn't show until the animal is 18 months to 6 years old. The calves can be infected by faeces or contamination of equipment, or through the colostrum. In older animals, milk yield drops, and chronic diarrhoea and weight loss occur. The animal does not recover.

Milk and blood samples will prove if you have got the disease in the herd. National Milk Records herds have a Johne's screening programme. If your herd is free from the disease, try your utmost to keep it so. Buying in new animals can bring it on to your farm without you knowing it. If the disease is in the herd, take away new born calves from their mothers at birth and feed with artificial colostrum, and don't keep calves from known infected mothers. You must cull diseased animals.

Joint Ill/Navel Ill

This disease affects calves that are usually less than 1 week old. Bacteria enter the animal through the umbilical cord soon after the calf has been born. Navel ill is when the navel becomes infected; it does not dry, and sometimes an abscess may form. In joint ill, the bacteria travel in the bloodstream to the joints, making them swollen, stiff, painful and sometimes hot.

Make sure the calves drink plenty of colostrum. Treat with antibiotics plus pain killers. At birth, all navels should be smothered in iodine to prevent infection and keep calf pens disinfected between batches of calves and pens clean.

Lameness

This is a big problem in dairy herds today, and dairy farmers spend a lot of time, effort and money on this. Any cow that goes lame should have its foot examined within 24 hours. The disease is caused by digital dermatitis, foul in the foot, laminitis, sole ulcers, and white line disease. Feet should be looked after and not left to go overgrown. Keep cattle's feet as clean as possible, and avoid letting them onto dirty, muddy and stony areas.

Digital Dermatitis

This is a very common cause of lameness indeed, especially on the skin around the heel and between the claws of the foot. The skin becomes inflamed with lesions and is very painful. To treat, use antibiotic sprays. Where cattle are housed, a foot bath with a disinfectant solution will help prevent the disease.

Foul in the Foot

In this disease, the skin is affected between the claws, and the bacteria only enter through the skin when it is damaged. Animals become lame, the foot swells, the skin swells between the feet and the digits and pus may be produced. There may be an increase in temperature. Treat with antibiotics along with anti-inflammatory injections. Control measures include improving gateways and walkways to avoid cuts between the claws.

Laminitis

This is a metabolic disorder and can be acute or chronic.

Acute Laminitis

Animals are in pain when standing or walking and the disease is associated with other problems, e.g. toxic mastitis and acid conditions in the rumen.

Chronic Laminitis

In this disease, there will be damage to the walking surface of the claws and the sole of the foot where bleeding into the horn of the sole may be seen. It is found where cattle are given high-protein feed, so this should be cut down to a lower level. Anti-inflammatory therapy will help. An expert foot trimmer can relieve pain by transferring the weight from the damaged claw to the healthy claw.

Sole Ulcers

These may develop after calving. Ulcers are found in the sole of the foot, as the name suggests, especially in the outer claw of the hind feet. An expert foot trimmer will benefit the animal, as will placement of a shoe block on the unaffected claw.

White Line Disease

The white line is a point of weakness where the horn from two different sources (wall horn and sole horn) join. Dirt becomes impacted, causing infection and abscesses making the animal lame. You can remove quantities of hoof and open up the abscesses. The foot can be bandaged and poulticed to draw the pus; however, bandages get wet and dirty, so you need to keep the floors clean. Shoe blocks manufactured for cattle can also help.

Leptospirosis

Leptospirosis causes abortion, infertility and poor milk yield. It is caused by bacteria, it can also affect humans, and a good reason for pasteurising milk is that it destroys all leptospire organisms.

Cows can be infected from contact with urine that is infected or by products of abortion. The bacteria like to live in water, so unfenced streams may bring in the disease from other farms. The animals will have a raised temperature and will lose appetite. The milk becomes very thick, similar to colostrum, and the udder becomes 'flabby' leading to the description 'flabby bag syndrome.' Abortion occurs 6–12 weeks after the first infection.

Antibiotics help the animals to recover, but they will not regain their full milk production. They may be immune for a time, but they will be susceptible to further attacks. Sheep can carry and excrete the disease, so mixed grazing is not advised. Some farmer friends in my area vaccinate against it.

Liver Fluke

Cattle are less susceptible to the disease than sheep. This fluke is a parasite whose life cycle employs an intermediate snail host, *Fasciola hepatica*, which likes moist warm conditions in the fields. In well-drained pasture that dries out fully, the snails cannot survive, and the fluke cannot live on.

Symptoms include reduced milk yield, weight loss and diarrhoea. Beef cattle will again lose weight and will not be doing well. Their livers are damaged by flukes, and heavy infestations of the liver can occur, which can cause anaemia and death.

Different flukicides can be purchased – make sure you adhere to the withdrawal periods. After using some of them, cattle cannot be slaughtered for human consumption for 60 days. Many flukicides are available, and withdrawal times vary.

Mastitis

Mastitis is caused by bacteria and is characterised by inflammation of the udder tissue. Dirty bedding and unhygienic milking conditions are a minefield for bacteria, which may infect the cows. It can also be caused by physical damage to the teats or udder. The bacteria enter the udder via the teat canal. They multiply very quickly indeed, producing toxins that cause inflammation. The cow releases leucocytes (white blood cells), which make the milk curdled, watery or even blood-streaked. Wash teats before milking, and if washed, thorough drying is essential. Teats should be dipped in a germicide after each milking. If a cow has mastitis, you milk her and then give her an intermammary infusion of antibiotic. The antibiotic is inside a plastic tube, which is inserted into the teat canal. Milk must be discarded for a specified number of milkings after the antibiotic has been introduced. Always strip some foremilk out of each teat before putting on the milking machine. This should be done in a special strip cup and not be squirted onto the floor.

Metritis

This is inflammation of the wall of the uterus caused by bacteria. It occurs soon after calving (usually in the first few days) but can occur in the first and second week after calving, sometimes seen after a calving problem when bacteria have entered the animal.

Retention of foetal membranes doesn't help the problem. Cows will suffer discharge from the vagina and become ill. You need veterinary advice on what antibiotic to use.

Milk Fever

Milk fever (hypocalcaemia) is a metabolic complaint caused by low blood calcium levels. Fat, high-yielding, older cows are susceptible, and the disease often appears within 1 day of calving. Sadly, the Jersey breed is prone to it. Cows will stagger about, lie flat and, if not treated, fall into a coma and die. To treat, give 300–600 ml of 40% solution of calcium borogluconate. Inject in different places under the skin, or if you are experienced inject into a vein (this can kill if you don't do it properly – if given too fast). Injecting into a vein gives a much quicker response.

Neosporosis

This is the most common cause of abortion in cattle. It is caused by protozoa – *Neospora caninum*. There are no clinical signs seen in the cow, but the cow passes the infection on to its calf while it is still in the womb. The cow aborts between 3 and 9 months of pregnancy (5–7 months being the most common), or she may give birth to a still born or premature calf. Cows cannot directly infect other cows.

The protozoa uses dogs or foxes as hosts. It becomes mature inside the dog and produces oocysts. The dog passes out the oocysts in its faeces. After 2 or 3 days, the oocysts become infectious. The cow eats the grass and oocysts, and they then migrate through the cow, infecting the central nervous system and causing infection and abortion. Dogs can take in the parasite by eating raw meat or afterbirth and then become contaminated. Blood tests will show if your cow tests positive for the disease. There is no treatment or vaccine available, so cull infected cows; the meat can still be sold. It is important to pick up all dog faeces on the farm and remove placental membranes and aborted and dead calves.

New Forest Eye or Pink Eye

Bacteria are spread by flies during the summer, infecting the eyes. The eyes become cloudy and pink, and tear stains will show under the eyes. Corneal ulceration may occur, and the animal may close its eyes to reduce the pain; sight may be lost on a temporary basis, but if left untreated the animal could lose its sight permanently. Use an antibiotic cream in the eyes or a long-acting systemic injection.

Poisoning

Lead

Lead can be found in batteries, grease, crank case oils and old lead piping, and if left lying around in fields and buildings, it is a hazard for cattle. Years ago, paint contained lead, and calf pens painted with this often poisoned calves as they licked and sucked it; fortunately, lead-based paint is now very rare. However, on old buildings, lead paint may be three or four layers down and be covered with modern paint – and so the cattle may find it.

Most cattle are just found dead, but signs of poisoning could include grinding of teeth and twitching of the eyes and ears. They become blind and press their heads into the corner of buildings. They will stagger and have convulsions, and will probably die.

Once you have observed the signs, it is still worth treating if the animal is not having convulsions. You could try vitamin B_1 injections and giving Epsom salts orally, or slow intravenous injections of a 5% solution of sodium calcium edetate. This is the most effective treatment. To prevent it, make sure you have no clutter lying about your fields and buildings – keep things tidy.

Ragwort

Ragwort now seems to be almost everywhere in this area – certainly many grass verges are full of it, and council workers at one time did pull it up, but it seems it is now just left to grow wild. I have seen a couple of plants on the grass verge quite near to our entrance gate, but I pulled these up well before they went to seed, as I certainly didn't want them spreading on to our farm land. I hate to see it growing in fields.

Ragwort poisoning usually occurs when the animals eat ragwort in hay and silage. The cattle become dull, lose weight, and have diarrhoea and jaundice; fluid collects under the brisket and jaw. The liver becomes damaged and actually shrinks in size.

There is no treatment. If you find ragwort growing on your farm – don't spray it with weed-killer and then put animals into the field soon afterwards; this makes it palatable as it is dying. Cattle usually won't eat it if 'fresh.' Plants are best pulled up when you only have a few.

Yew

The yew tree, *Taxus baccata*, has been traditionally grown in church yards for hundreds of years. The tree is very poisonous except the bright red aril surrounding the seed. The major toxin is the alkaloid taxine. Death may follow within a few hours of ingestion. If symptoms are seen, they include accelerated heart rate, convulsions, staggering, difficulty breathing and heart failure.

Dumped yew clippings in a field from someone who has clipped their hedge will kill the animals, and yet the person responsible will not know what they have done. Some trees are much more toxic than others, but you can't tell which. There is no treatment.

Pneumonia

Pneumonia is common in housed calves between 1 and 5 months old causing lung damage, slow growth and sometimes death. Cattle will have a raised temperature and discharge from the nose, and will cough, be looking dull and have raised breathing in addition to eating less. Pneumonia can be caused by both viruses and bacteria, and is spread through the air in droplets, from the infected calf to other calves. This is why the calf house should be well ventilated and draught-free and also why you should try not to mix different batches of calves. Animals are susceptible when stressed, so treat calves as quietly as possible when carrying out stockman's tasks. Bacterial infections can be treated with antibiotics, and vaccines such as Rispoval are available for protection from the disease.

Ringworm

Ringworm is a fungal disease that infects the skin and is most often seen in younger animals. The fungus can produce many spores that can live in the calf house for years. The crusty skin lesions can be found mainly on the head and neck in circular patches. They may stay on the animal for months and then heal over without treatment. It is more common during the winter with cattle kept indoors.

Ringworm.

Ringworm is an infectious, transmissible disease that looks unsightly. It can be transferred to humans. I've got the scars on my leg to prove it – they show up red when I'm in the bath! It was called 'dairyman's itch' – an expression that is little used today – and can be treated with topical anti-fungal treatments, but many farmers do not treat it at all, as it will clear up on its own, especially when cattle are turned out to grass.

Salmonella

Salmonella dublin usually causes Salmonella in calves and older cattle. Sources of infection include faeces from infected cows and probably birds, dogs and cats. It can be present in cattle without them showing signs of the disease. The cattle, however, may have diarrhoea, which can be bloody, and fever; milk yield may drop, cattle will look unwell, and abortion can occur.

Treated cattle may become carriers, spreading the disease from their faeces without themselves showing symptoms. Once infected many will die if not treated. Treat with antibiotics. Calves that have not had enough colostrum are at risk and, if they are not drinking fluids, these must be replaced with an approved solution, consisting of electrolytes and energy. Salts or electrolytes must be replaced to keep the electrolyte concentration of the body fluid constant. Separate infected cattle and keep buildings clean. Vaccines are available.

Scours

Scours or diarrhoea can be a real problem with your calves. It is more common in calves reared artificially than in suckled calves, and young calves are more prone because of their liquid diet. A calf that has drunk plenty of colostrum is less susceptible to scours. Loss of fluid from the calf can cause dehydration. Scours can be caused by a virus, bacteria or protozoa. Calves will be lethargic with head and ears down, have watery faeces and may be dehydrated – if you pinch the skin, it stays tented. The calf may have a fever or cold extremities and may not drink. Vaccines and/or colostral supplements may help to prevent outbreaks. The best line of defence against scours is to ensure adequate colostrum intake in the first few hours after birth.

To treat – administer a veterinary approved solution, which will consist of electrolytes and energy. Electrolyte is a medical scientific term for salts, specifically ions, and during scouring the calf will have lost some of these salts. These salts or electrolytes must be replaced to keep the electrolyte concentrations of the body fluid constant. Without replacing the fluids and electrolytes, the calf will dehydrate.

Tetanus

This is a clostridial disease that can affect all ages of cattle. The bacterium *Clostridium tetani* is found in the soil. The bacteria get into a wound and reproduce, causing infection.

The bacteria produce toxins that spread to the brain, which is fatal. Symptoms include the animals not wanting to move, becoming stiff, the third eye lid coming across the eye, lock jaw, bloat, respiratory failure and collapse. Affected animals are best left in a quiet area and can be treated with penicillin. Vaccination is available.

Tuberculosis

This is an infectious disease of cattle that can be spread from cow to cow by the bacterium *Mycobacterium bovis* (M. bovis). Chapter 7 explains TB testing. The bacterium can survive in buildings and in faeces, and is spread by droplet infection, by infected faeces or in the milk.

Symptoms vary: animals may have a chronic cough, lose a lot of weight and produce clotted milk. Animals are not treated but slaughtered.

Worms - Parasitic Gastroenteritis (PGE Gut Worms)

Worms are parasites that are found in the stomach and intestines, causing digestive problems. Animals will lose weight, have diarrhoea and lose appetite, and most animals in the same group will become infected. Eggs are passed out in the faeces and need some time to live on the pasture. Larvae hatch from the eggs and are eaten by the cattle, and develop into an adult inside the animal.

Lungworms

Lungworms live in the air passages in the lungs and can cause coughing; animals will lose weight, stand with their head down and not want to move. Animals showing signs may die quickly. Vaccinations plus various worming products are available – drugs and slow-release boluses.

Schmallenberg Virus

This is a new disease in this country that affects sheep and cattle. It is a virus that causes mild illness in stock, but the main problem is that it causes stillbirths, abortion and foetal abnormalities. It is not a notifiable disease, but affected farms have been requested to report cases to their veterinary surgeon. There is no treatment, but a vaccine is now available.

Swallowed Wire or Other Metal Objects

Wire or other metal objects may be swallowed when the animals are grazing or feeding indoors. This leads to stomach pain. The wire may find its way to the reticulum (the second stomach) where it could penetrate the wall and then enter the peritoneal cavity, causing peritonitis to set in.

The animal will have a raised temperature, and stand with its body arched; it will not want to eat and will grunt. It will need antibiotics, and the wire will need to be removed by surgery – or they can be given a magnet, which will hopefully bring the wire back into the reticulum. Many farmers place a magnet in every animal's rumen, and so if a wire is swallowed, the magnet is already there.

Chapter 12

Making a Start with Sheep

Mary had a little lamb
She tied it to a pylon
10,000 volts shot up its tail
And turned its wool to nylon.

My next top tip to get into farming is to start with sheep. To start right at the bottom of the farming ladder with a tight budget and a full-time job is to first find some grass keep. Grass keep means you put the sheep on the land, and you are just paying for the grass and have no other rights.

Sheep are kept for meat and wool, but the price of lamb should hopefully make you a profit, whereas the price of wool is poor. Sheep are also kept to sell as breeding stock. However, not all of your sheep will be good enough to sell as breeding stock, and some will end up as meat. Lamb is very popular in this country, and the term is used to describe the meat from sheep that are up to a year old with no permanent incisor teeth. A young sheep over 1 year old is called a hogget, and mutton is the meat of older rams and ewes. Mutton is much less popular than lamb and has a richer, stronger flavour but can be tough. Some sheep are kept for milk production, and some of this milk can be made into cheese and yoghurt.

Sheep can be bred on the farm (as I do) or can be bought as weaned lambs or adult sheep. It is much easier to find grass keep than to find land to rent on a permanent basis. You can buy some store lambs and then keep them until they are fit for slaughter – you might buy them in August or September and then sell them in November, December or January. The most important thing is that the grass keep should be yours to put the sheep on until they are ready for market, but if the worst comes to the worst, and you have a few lambs left that are not ready and you have to give up your grass keep, you can sell them as stores again. They should have grown since you bought them. Another problem could arise if it comes a lot of wet weather, and the lambs make a mess in the field – the field becomes poached and very muddy. One of my friends kept his sheep on grass keep. It was a local dairy farm, and the landlord asked him – in fact told him – to remove the lambs immediately because the field had become very wet and muddy. This grass keep was just a word-of-mouth agreement, and nothing was in writing. The lambs were removed the following day; luckily my friend had grass keep elsewhere and so did not have to sell them immediately.

Hopefully you will have enough grass for your stores, but if a lot of winter weather comes, you may have to feed hay, a root crop such as fodder beet or some concentrates. You will probably get away with having no farm buildings for this – your first enterprise – but you will need some hurdles to pen the sheep in for sorting them out, treating feet and worming. If you can't afford hurdles, you can create makeshift pens out of round posts knocked in the ground and then fasten on some pallets. This doesn't look as good as the hurdles but it can work – fortunately I haven't had to use pallets.

Texel cross Lleyn store lambs.

If the farm is not fenced well enough for sheep, which is often the case on dairy farms, you will need rolls of electric netting and units to power your netting. You will also need to buy some foot shears, dagging shears, a worming gun, syringes and needles, some marker spray, a product to control blowfly and some wormer. In my early sheep farming days – before I bought my first field – I took on grass keep, fencing it with electric netting and travelling every day to check on the stock and finding them out when I got there! You must learn how to catch a sheep and 'turn' it (sitting it on its backside) so that you can take care of its feet.

You can buy big, strong store lambs that will take just a matter of weeks to finish or you can buy some smaller, cheaper lambs, which will take 6 months or more to finish. If you are buying store lambs that you intend finishing for the butcher, go for lambs sired by a terminal sire such as a Suffolk, Texel or Charollais, as these sires breed good strong lambs for the table. There is a real skill in buying store lambs. The lambs should look the part and be healthy. Of course, you will want to pay as little money as possible for them, but if you buy cade (orphan) lambs or thin little things, they may be cheap, but you only have to have one or two die, and that is probably your profit gone! These little lambs may take a lot of finishing and will probably need concentrates to get them good enough for the butcher. I would sooner buy the stronger lambs, which will be more expensive, but they are less likely to die and won't take as much finishing.

If you have not bought lambs before (and you can buy privately or in a market), take an experienced farmer with you, if you know one. Profit per lamb will certainly vary from year to year – be aware that you could lose money; lamb prices can go up or down in the market. A stock broker gives the same advice to his clients when you buy stocks and shares – they could go up, or the price could come down. Adam, my accountant, came out to the farm to see me to go through the books. 'How are you?' I asked. 'Quite good thank you but I can't get to sleep at night very easily' he replied. 'Have you tried counting sheep?' I asked. 'Yes, that's my problem. I make a mistake and then I'm awake for three hours trying to find it!'

To make money out of your lambs, don't sell them at market or privately to a butcher. Sell them as whole lambs or half lambs for freezers. You should ask people at work or your neighbours or relations 'Do you have a freezer?.' Then you ask, 'Do you eat lamb?' If they say yes, you will have written down on paper the cost of leg, shoulder and chops in the supermarket (you will have been to the supermarket with a pen and paper previously). You then tell them how much a whole lamb would cost from you – a saving for them and much more money for you because you will have the farmer's profit and the butcher's profit. You can advertise them as 'English lamb, locally reared.' It can be a bit risky if someone orders one and doesn't turn up for it; you are left with it, so leave room in your freezer. If you do this, make sure you take a deposit from each customer if you don't know them. You may want to set up a website to sell your lamb.

Lambs ready for the butcher being sold at market. My Texel crosses are in the pen in the foreground.

You will need to book your lambs in at an abattoir in advance. You need to make sure they will butcher them for you –

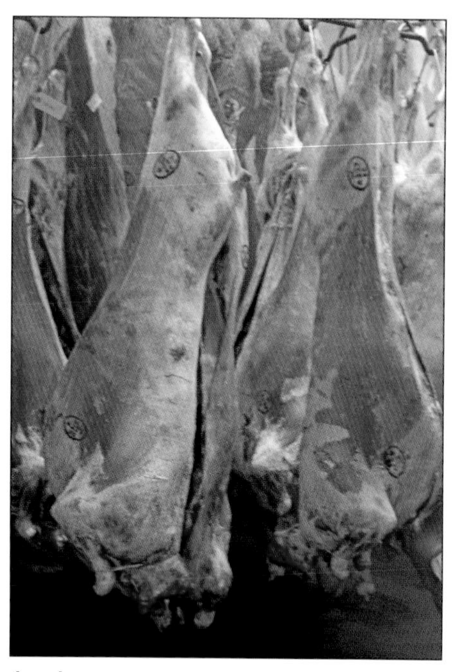

Lamb carcasses.

they usually cut them up and put each joint into a plastic bag and as many chops as you want in each bag. Some of your customers will want two to a bag; others four or more, depending on the size of the family. You can have the legs left whole or cut into two – again depending on the size of the family. I have sold lambs this way for about 33 years and have made more money this way than selling to a market or a butcher. However, if you can't find enough customers for lambs for freezers, the surplus will have to be sold at market or direct to the butcher. Most of my lambs are sold for pedigree stock, or ewe lambs kept here for breeding, but lambs with poor markings or other problems will end up as meat.

Another method of getting on the sheep-farming ladder with just grass keep and no buildings is to buy in barren ewes; these will be sold any time after the lambing ends or later on in early autumn. Ewes will be culled from the breeding flock; they may only have one quarter (only one side of the udder producing milk) or a few or no teeth, and they may be quite thin, as they have worked hard rearing their lambs. These sheep would need good pasture and would be sold as mutton when they had put on weight, hopefully making you a profit. Be careful if you intend to by old ewes at a livestock auction. Butchers buying at auction often do not like newcomers, and they will run the bidding up, and so you will end up paying more for them than you bargained for. You probably need to find a professional buyer to bid for you, and then you pay him a fee – which will be less than being 'run up.' One of our neighbours a few years ago did very well buying in about 200 barren ewes. The price for these cull ewes went up in the winter when he sold them in March, and he made a very good profit. There seems to be a shortage of old ewes in this area during March. Ewes that have been scanned to see if they are in lamb will probably have already been sold. Sheep farmers are busy lambing at this time of year and don't have time to take old ewes to market. However, because they are old, you may have some die, and if too many die then your profit margin may disappear. Of course, the price of cull ewes could go down instead of up. A few of your cull ewes might actually produce lambs! I have never bought in old ewes and prefer to keep a breeding flock.

You can have grass keep, but often there are no buildings on it. You can rent a field without buildings and rent buildings on another farm. Of course, it is much better if you can buy your field and then put your own buildings on it or even better if you can buy the lot straight out.

To keep a breeding flock of sheep, you need some sort of building. I know some people manage with straw bale shelters and tin sheets, but these have to be constructed well, or they could be easily knocked down or blown apart in a gale. Keeping a breeding flock of sheep is a good way to farm – you can put your life and soul into it, and reap the rewards. Pedigree breeding for me is very rewarding and satisfying. You can start small with just a few hectares and just a few sheep. The number of ewes kept per hectare depends on where you are situated. If the land is very poor, hilly or mountainous, you can keep fewer sheep than on lush lowland pasture.

The quality of the grass, the use of fertiliser and whether or not you can get the sheep off the land in the winter will all affect the number of ewes and lambs kept. An average would be 12 per hectare or 5 per acre, but with fabulous grass and fertiliser, this could perhaps rise to 20 per hectare or 8 per acre.

Grass needs a rest, which gives it a chance to recover and grow, so you rotate your sheep around the fields. If you only have one field, it is best to split it up into two or three smaller fields with fencing posts and sheep netting. You can use electric netting to split the fields, but I have found this unsatisfactory, as some of my Kerry Hill sheep would jump the fence, and if the battery is low, they may get out. When I owned just the one field, I split it into two, and it remains split to this day.

A big ewe with twin lambs will eat more grass than a small ewe with a single lamb. Most farmers in my area keep sheep to produce lamb for the table. They use a terminal sire – that is, a ram that produces meaty, well-fleshed lambs in a short time – breeds such as Suffolk, Texel or Charollais. These are used on ewes that are hardy, prolific and good mothers, and produce plenty of milk for their lambs, such as mules (a North of England mule is a cross between a Bluefaced Leicester ram and a Swaledale ewe).

Sheep are usually sold in the early autumn – the start of the shepherd's year. You can buy a ram lamb to breed from – he will serve if he is mature enough. A shearling is better because he is bigger and can probably manage to serve more ewes. I have bought shearlings which have not yet sired any lambs, but I have also bought 3- or 4-year-old rams that are proven sires. I have visited the farm and looked at the lambs he has sired, which is a great advantage. With an unproven ram, you run the risk of him being infertile, or he may produce lambs with a fault, e.g. siring lambs with black patches of wool in their necks; this can certainly happen with the Kerry Hill breed. A 3- or 4-year-old ram would probably be less expensive than a shearling, but it's up to you.

When buying females, you can buy ewe lambs in the autumn – that is, lambs born in the spring about 6–8 months old. These can be put to the ram, and many farmers breed from ewe lambs. I don't, as these sheep are still growing, and sometimes they don't grow into such big sheep the next year. They can also make poor mothers. I like to put ewes to the ram for the first time when they are 18 months old – these are called shearlings. Shearlings will be best to buy, but you can also buy in older ewes that are more experienced at lambing and are good mothers, but beware – some farmers will get rid of problem sheep because they have had trouble with them, and they don't want the same trouble again, e.g. ewes that don't produce enough milk, ewes that have prolapsed or ewes that are poor mothers. I buy ewe lambs and keep them until they are 18 months old before putting them to the ram, or I buy shearlings, but I don't buy many females because I prefer to breed my own.

You can also buy old ewes to breed from. You must not buy ewes with only one quarter. If only one side is working, the ewe will not be able to feed two lambs. Broken-mouthed ewes, i.e. ewes with only a few teeth or no teeth, will find it difficult to eat enough grass to produce the milk required to rear a lamb properly. I would certainly not buy old ewes to breed from. You can also buy ewes in the spring that have lambed, and so you buy the ewes and lambs all together. The ewes could be shearlings or older ewes. Older ewes may be cheaper, but they won't live for as many years. You can sell the lambs in the autumn as stores or finished lambs and sell the ewes or keep them.

When buying rams, you must feel to see if he has two large testicles – the same size. Both rams and ewes should be able to walk well and have good strong bones with legs straight and not bowing out or inwards. They should be good on their feet and up on their pasterns (the joint above the foot). If the sheep is down on its pasterns and walking on its heels, this may affect the performance of rams and ewes.

All sheep should be 'good in the mouth,' which is an expression we use to describe the incisor teeth. The bottom teeth should meet the pad on the top jaw in the correct position, not be undershot or overshot.

I prefer to pick out and buy the bigger sheep out of a flock that I am choosing from; usually we say in showing 'A good big one will beat a good small one.' A big ewe with a long level back and not dipping behind the shoulder with plenty of heart room and a good spring of rib should hopefully do well. They should also have good tight wool and, if pedigree, have the necessary breed points to be admitted into the flock book.

The Sheep Year Here at Oak Tree Farm

The sheep year starts in the autumn – we flush the ewes in the early autumn, which means putting them on poor pasture so they lose a little condition, then put them on rich pasture to get them into good breeding condition, which will hopefully produce more twins.

Ewes are put to the ram in the autumn because the ewes will come on heat (oestrus) for 24–36 hours and, if not in lamb, will come on heat 16–17 days later. One ram in with 50 ewes is plenty, and a ram lamb should be given less. Our Kerry ram at the moment has about 20 ewes and the Derbyshire Gritstone ram about 30 ewes. It is good stockmanship to change the rams – so when one of our rams has been in and served the ewes, we take him out and put another one in just in case the first one was infertile. I raddle the rams using coloured crayons. A raddle is a harness that holds a pad of coloured wax called a crayon. The crayon sits between the front legs of the ram. (Where does the ram get his raddle from? At a Tupperware party of course!) When he mates with the ewes, he will leave a coloured mark on their rumps. If you change the colour of the crayon every 16 days, you can see batches of ewes with the same colour that will lamb together as a group. You will find some ewes will be more difficult to get in lamb and may have two or three colours on their rumps. The freshest last colour is the one to put in your diary. Hopefully she will lamb 147 days later, or of course she could be barren. The number 147 is the highest break in snooker, so if you are a snooker fan it is an easy number to remember – incidentally, I am not a snooker fan. Both the Kerry Hill and Derbyshire Gritstone rams go into our ewes in October, and so lambing takes place in February and March. If I was just keeping sheep for meat production, I would put the rams in a lot later and then lamb the ewes at the end of April; then they could be turned out with their lambs very soon after lambing. I like these pedigree sheep to lamb fairly early to have some good, big, strong lambs to sell as pedigrees in the autumn. Ewes are scanned between 70 and 105 days of pregnancy, the advantage being that this will identify ewes that are carrying singles or multiples, and will show which ewes are barren. A ewe carrying twins or triplets will need more feed than a ewe carrying a single lamb.

A farmer approached his veterinary surgeon and said 'I've bought another ram, and I am so annoyed because he is the same as the last one I bought. He is just not interested in serving the ewes.' 'I'm sorry to hear that,' replied the vet. 'Last time you gave me some tablets, which were marvellous. After taking them, the ram was rampant and served 50 ewes in four days.' 'I can't remember any tablets. Are you sure?' asked the vet. 'Yes I'm sure. They are pale green and taste like peppermint!' replied the farmer.

Sheep are vaccinated against the clostridial diseases and also against pneumonia with Heptavac P. Ewes are injected twice in the autumn, and then 4–6 weeks before lambing; antibodies in the colostrum are passed on to the newborn lambs, preventing disease.

Ewes can be left out all winter and can lamb outside, but I prefer to bring them indoors in January, lambing them inside and then turning them out with their lambs when the weather is kinder, and some grass is growing. Lambing is better for me indoors, especially on a cold wet or snowy night in late February or early March. If the sheep are brought indoors early, they will just need hay to eat (some farmers feed silage or straw, but I much prefer hay). Big bales of hay are now more common than small conventional bales. These big bales are difficult to handle without a telescopic forklift, but it is getting harder to find a contractor with a conventional baler. Large farms want large bales, and farmers don't want to be lifting small bales anymore, but if you only have a few sheep, small bales are best. I have tried larger bales but have gone back to conventional bales.

Seventy per cent of the foetal growth occurs during the last 4–6 weeks of pregnancy, and rumen space decreases with the growth of the lambs, so we start to feed concentrate 6–8 weeks before lambing in two feeds per day. You can mix up your own ration with ingredients such as rolled oats, sugar beet pulp and some sheep pellets, or you can buy a balanced ration in the form of sheep pellets, which will be 16 or 18% protein. We start with a small amount of food and work the ration up to perhaps 0.5 kg or 1 kg per day depending on the size and breed of the sheep, and also if she is carrying a single or multiple lambs.

I used to mix my own ration but now use the easier option of buying in sheep pellets or sheep nuts. This saves the bother of mixing. Sheep feed bought in bulk will be cheaper than in bags. We make sure the sheep have plenty of trough space – they should all be able to get their mouths in to feed together.

My ewes enjoy good quality home made hay, however, they will eat less in late pregnancy, and so we feed concentrates to keep them healthy. They will waste hay given half a chance by just eating the best and pulling the rest out of the hay racks onto the floor. They need clean water at all times, as they will drink more indoors if they are off grass, and of course we use straw for bedding.

Lambing

I get organised – not leaving things to the last minute; lambs could be born early, and with no equipment the lambs could die. I make sure we have got the following:

1. Colostrum supplement. This is top of my list because if you have a lamb born prematurely in the middle of the night, and the ewe has got no colostrum, you need to feed your lamb with a bottle. If it is small and weak, it may die without immediate nourishment. You can also feed 50% dextrose for weak lambs. After a couple of days, we stop feeding colostrum and start them on lamb milk replacer.
2. Lamb bottles and teats and a feeding tube. It is no good having the colostrum and being unable to get it into your lamb. A stomach tube is a tube placed down the throat into the stomach, and the colostrum is poured in. This is used when the lamb hasn't got the strength to suck. Once the tube is inside the lamb, I listen at the end of the tube by placing it to my ear to make sure it is not in the lungs (if it is in the lung it sounds like deep breathing, if it is in the stomach you may hear only a slight gurgle).
3. Lambing lubricant. This is put around the ewe's vulva and into the vagina, and on your hands if you get a difficult lambing. Some lambing seasons we get away without a difficult lambing.
4. Nylon rope. Again, this it to assist with difficult lambing.
5. Iodine to spray onto the lambs' navels.
6. Ear tags plus the applicator to put them in with (look after your applicator because it will last for years).
7. Castration rings. These are used to castrate unwanted ram lambs and also for docking tails. You will also need the tool to apply the rings.
8. Marker spray to spray identification numbers on the ewes and lambs.
9. Needles and syringes to inject antibiotics plus a larger oral dosing syringe.
10. Antibiotics.
11. Bottle of injectable calcium.
12. Propylene glycol and electrolyte solution in case of pregnancy toxaemia.
13. If a ewe prolapses, I can fit a vaginal prolapse retainer ewe spoon to hold the prolapse back in, but a suture made of nylon tape is best. I have only had a few ewes prolapse – thankfully.
14. Disinfectant.
15. Heat lamp for weak lambs to get extra warmth. Very cold lambs are brought into the bungalow to be warmed up and then given colostrum.
16. Enough lambing pen sides or hurdles to make individual lambing pens.

17. Enough water buckets.
18. A list of ewes' ear numbers to record the lambs. Records of pedigree sheep are needed by you and various breed societies; they can then be given a unique individual number.
19. A diary to record work completed.

Before you lamb your own flock, you should get experience working on another farm or attend a course at a college that teaches agriculture. I have had newcomers spend time with me, and some have gone on to have much bigger flocks than me and made a tremendous success of their flocks.

I check on the ewes as often as possible, and I start about a fortnight before lambing begins, as some may lamb early – by this time, many will have bagged up, which means that the ewe has started to develop an udder.

Different farmers or shepherds have their own way of doing things and check the sheep at different times especially in the evenings, nights and early mornings. I have tried 9 pm, 11 pm to 12 midnight, 3 am and 6 am, and for years would get up automatically at 3 am – the middle of the night! Now I am older, I check at about 6 pm, 8 pm, 11.30 pm and again between 5 am and 5.30 am. If you have got a ewe lambing, stay with her, but on the coldest of nights, it is a pleasure to look and see nothing is happening!

Another scenario is you look at them at 11.30 pm – nothing is lambing, but you think one might lamb soon, so you can go to bed for an hour or you can read or watch television. A big disadvantage of lambing in the night is you are often very cold when you get back into bed, and so it takes longer to get to sleep. With a large flock of sheep, you will have more staff, and you can have day shifts and night shifts, which means someone can be present with the ewes all the time.

A ewe about to give birth in a field will usually leave the rest of the ewes and will try to find a quiet spot to give birth. Indoors, she can't leave the rest of the flock, so she may get into a corner or lamb down the side of the building. The next sign that she is giving birth is when she starts to paw the ground; she will then lie down, stand up again, walk round in a circle, lie down, stand up. You can either leave her to get on with it or watch discreetly. Lambing will usually be fairly quick at this stage. It might be half an hour or less or a few hours – sheep are all different like most mothers to be. She will start to strain, pointing her head skywards, and the water bag will appear. Most lambs are born in normal presentation – the lamb will come head first with the front feet next, the ewe will continue to strain and hopefully the lamb will be born without help from the shepherd. However, problems do occur, and the ewe may find it difficult to give birth. Problems could include a breached lamb – a lamb coming backwards – but often

Lambing at Oak Tree Farm 2012 – Jonathan lambs his first ewe.

these are born without help, head out first with no legs showing or one leg back or two lambs coming together. With experience, you soon spot the problem and deal with it.

The head is the difficult part. The head follows the front legs out, and once the head is out, the rest should be easy. Lambs are born with mucous membranes covering their faces; ewes will lick this away quickly and instinctively so that the lamb can breathe – it is very important that this is removed. I, like many other farmers, have lost lambs because the ewe has not been quick enough to remove this membrane. I am always quick to get in the pen and remove this membrane, and if I am there I will certainly remove it before the ewe has a chance to, just in case.

If the lamb has been born and is not breathing, you must work very quickly indeed. There are various methods I use to get it going – moving its front legs vigorously or holding it by its back legs and swinging it or poking some straw up its nose, or doing all of these things. Once breathing, I move it round to the front of the ewe so that she can lick it. I stand the ewe up if she hasn't got up. If she starts to lick the lamb, I know that she will mother it. If she doesn't lick it immediately, she may do, it's just that the penny hasn't dropped. This sometimes happens with first-time lambers. If she doesn't lick it after a few minutes, I lift the lamb up to her head height and put it next to her nose. If she still doesn't lick it, wet your fingers with the fluid on the lamb and put your fingers in her mouth or you can put the lamb's tail in her mouth because that will be very wet. Once she has got the taste, hopefully she will mother it.

I then need to check to see if the ewe has got colostrum (colostrum is the first milk containing the beneficial antibodies). Milk some out from both quarters – most ewes will have some, but sometimes it can be late coming. If she has only got one side working, she will probably only rear one lamb well or two lambs not very well. Once the lambs are born, the ewe and her lambs are placed in an individual lambing pen. To get her into the lambing pen, we pick up the lambs and walk backwards slowly, keeping them close to the ewe. Make noises like a lamb, and the ewe should follow – a good mother will follow 'like a train' and be very concerned. However, we get some that won't follow you, and we have to catch them and move them in with some pushing and shoving. We dip or spray the lamb's navels with iodine as soon as possible after birth to stop infection – especially joint ill or navel ill. Infection enters the body via the navel cord causing swollen limb joints, and lambs could die. Clean straw in the pens is also essential. We don't put the water bucket into the pen until we are sure she has finished lambing because she may give birth into the bucket, and the lamb could

Ewe and lambs in lambing pen or mothering up pen.

drown. However, we put the water in as soon as the last lamb has been born because I can guarantee she will have a long drink next. We keep them in the individual lambing pen (which should be large enough for the ewe and lambs to lie down and turn around easily) for at least 24 hours – 48 hours is better – and longer if you have a problem with either the ewe or lambs.

When you foster a lamb, always take away the strongest lamb from a set of twins or triplets. If you have a ewe give birth to a single dead lamb, she really needs to work for her keep by rearing a lamb. Work as quickly as you can after she has had a dead one. I always put my hand inside her and pretend I am pulling out another lamb. I then bring the lamb I am hoping to adopt out from under my coat. I will have smothered this lamb in the birth fluids from the dead lamb to get as much smell as possible on to the live lamb – often this is successful. Alternatively, I have skinned dead lambs and put this 'coat' onto the live lamb with success and I have also used lamb musk – a product with a heavy smell that you spray onto the ewe's nose and onto the lamb – this has also been successful.

While the ewe and her lambs are in the lambing pen, we carry out the following jobs after 24 or 48 hours, all the jobs together in this order:

1. Check the ewes' feet to make sure they are not overgrown or have foot rot.
2. If you are going to castrate any lambs or dock tails, use rubber rings. I definitely don't do this job at under 24 hours old because I always think the lambs have got enough to cope with, such as finding the colostrum.
3. Ear tag. Copy the ear number of the ewe and the ear number of the newborn lamb into your record book. This is by far the easiest way to do it – if you wait until they are older and living together in a flock, it is more difficult to find them.
4. We spray a number on the side of the sheep and put the same number on her lambs. We use three colours: perhaps blue for a ewe with a single, red for twins and green for triplets. The spray will last longer and show up well on the lambs, but the ewe's number will be harder to read in a month's time, so we spray the ewes with more spray. Once we have sprayed numbers on, we know all the jobs have been done, so we don't spray until we are sure –we then move on to the next ewe and lambs.

We keep feeding concentrates twice a day and turn ewes and lambs out into small groups, keeping them indoors if the weather is bad, but when the weather improves and we have enough grass, we then turn them out into the field. We introduce the lambs to creep feed as soon as possible. They remain on creep feed for at least 6 weeks.

Shearing

Shearing.

Sheep are shorn once a year, usually in May and June or early July. Professional sheep shearers using machines work long hours and can shear a sheep very quickly indeed. If you want to shear your own sheep, college courses are available for you to attend. If you want to show some of your sheep, you will have to check with the breed flock book to see when you can shear. Derbyshire Gritstones are sheared after 1 April, but Kerry Hills are best sheared in late December or early January, if you are showing these breeds. The shearer shears so that the fleece comes off in one piece, at the same time making sure he doesn't pull the wool over her eyes! Colin shears our sheep every year, sometimes helped by Sam.

Management During the Summer and Autumn

The ewes and lambs are treated for worms about three times during the summer with an oral drench, and the fields are grazed in rotation. The sheep are also sprayed against blowfly twice, during the summer and in early autumn. The lambs are weaned during the second week in July and are put on to good pasture. The ewes are either kept indoors for a few days and fed on straw or put on poor pasture so that all their milk dries up. We check that their udders are fine and not bursting with milk twice during the next week and also check for mastitis. Ewes return to better pasture after a few weeks, and the lambs continue to graze. Lambs start to get lamb pellets in early autumn, pedigree ewe lambs and ram lambs are sold privately or at breed society sales during September and October. Lambs not good enough to sell as pedigree stock are either sold at market as finished lamb or sold privately for lambs for freezers. The best pedigree lambs, of course, are kept for breeding. The sheep's year soon passes, and it is time to flush the ewes again.

Showing

You may want to show your sheep after you have got established. You need to look at different breeds at agricultural shows. Choose a breed, talk to the exhibitors and then join the breed society. Certain breeds take a lot of work to get them ready for showing. Kerry Hill sheep need shearing at the end of December or the beginning of January. They need to be fed really well on concentrate feed all through the show season plus hay and of course fresh water. They need to be washed about a month before the first show, and then you must keep the wool looking good by carding and trimming. You can learn carding and trimming by watching other exhibitors and then practising. Carding combs consist of small metal spikes, which

A home-bred Kerry Hill ram lamb at the Royal Show with Jonathan aged 18 months.

Some of the 1600 rosettes won with the sheep.

are used to pull the wool up to get rid of knots. Trimming is a skilled job and takes time to learn. I suppose it is one of my specialities, but then again it has taken me years of 'shear' determination to get it right. The sheep need halter training. Tie them up at first using halters made from old socks. These halters will not cut into the sheep's face. Just leave them tied up for a short amount of time and then the next day a bit longer. Tie them up in a row and not on their own, but leave plenty of space between each sheep, as you don't want them tangled up. To lead them – walk an old experienced sheep in front, and the newcomers will follow – if they are stubborn, another person can walk behind them to encourage them even more. At the end of your walk, get them to stand still and square. One person needs to hold them on the halter, and another person becomes the judge – touching the sheep, looking in their mouths and, in the case of rams, feeling for two testicles.

Suffolks, Hampshire Downs and Southdowns will all want to be worked on. Easier breeds such as the Derbyshire Gritstone and the Lonk have their faces and legs washed and no carding or trimming done at all, which makes life much easier.

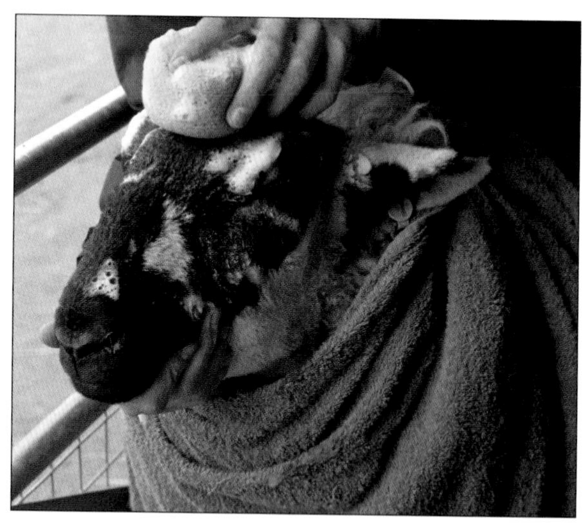

Washing a Derbyshire Gritstone's face ready for show. Note the towel wrapped around the neck to stop the wool getting wet.

You will have to watch sheep being shown, because there is a skill to it. I have spent 33 years showing sheep and many years judging at shows. I am proud to say I have judged Kerry Hill sheep at well over 20 agricultural shows including the Royal, the Royal Welsh, Castlewellan in Northern Ireland, Bath and West, East of England, Anglesey, Three Counties and the Shropshire and West Midland, and my only foreign appointment to Holland. It is a good hobby, and I am pleased our son is following in my footsteps.

Dairy Sheep

You need plenty of grass and good buildings including a parlour and milking machines. The British Sheep Dairy Association recommends a flock size minimum of 250–300 ewes, and they should have a lactation average of 300–400 litres (lambs are reared artificially).

Milking dairy sheep.

Breeds include the Friesland, Lacuone and British milk sheep, the Friesland being the most common. Sheep milk is sold fresh and frozen. Ice cream and cheese are also made. It is a specialised job, and I would suggest to the first-generation farmer – keep sheep for producing lamb or pedigree stock. However, if this is really what you want to do, get experience with sheep first, and then take plenty of advice.

Sheep Breeds

To produce finished lambs for the butcher or store lambs, many farmers keep crossbred ewes such as mules. They are the commonest crossbred ewes in Great Britain (about half the population). The sire, a Bluefaced Leicester, passes on the excellent quality of his meat and wool to the lambs. The ewes used to breed mules include the Swaledale, Scottish Blackface and Welsh Mountain. The ewes milk well, are good mothers, are hardy and live a long life. I recommend you start out with some mule ewes and put them to a terminal sire or start with easy pedigree breeds such as the Kerry Hill and Derbyshire Gritstone.

Mule ewes.

The sheep breeds in Great Britain can be separated into four different groups.

1. terminal sires;
2. crossing group;
3. wool group;
4. mountain and hill group.

There is also a breed which does not fit neatly into any of these groups – the Easy Care sheep – this is discussed further after the four main groups.

Terminal Sires (Down Breeds or Shortwools and Continentals)

These sheep need good conditions and will not like life at the top of mountains. They have a blocky appearance and are quick to mature into good-quality meat and wool. They produce terminal sires for crossing with other breeds and crossbred sheep – siring finished lambs for the butcher. As well as some British traditional down breeds, the group would also include many continental breeds. Of course, all the lambs born will not be terminal sires, half will be female, and many of these will be sold as breeding ewes. Pedigree lambs that are not good enough to breed from will end up as finished lambs for the butcher.

Suffolk

This is a handsome breed of sheep with no horns, found throughout the world. It evolved from crossing Norfolk Horn ewes with Southdown rams in Suffolk. It is certainly one of the oldest native breeds; the breed society was formed in 1886.

Suffolk ewes and lambs.

The pure-bred Suffolk breed is unmistakeable, with all black legs and all black head and ears, and the ears drooping down at right angles to the head. It has a really good fleece, which is used for various products, including knitting yarns and tweeds. This is a quick-growing breed, producing excellent, top-quality prime lamb. It is arguably the best British Down Breed.

Other Down breeds include the Dorset Down, Hampshire Down, Oxford Down, Ryeland, Shropshire and Southdown.

Texel

The Texel is a continental breed originating from the Island of Texel off the Dutch North-Sea Coast. This imported breed has become the most dominant terminal sire in Great Britain. It has a white face and white ears, sometimes with a few black spots on them, white legs and no wool on the head and legs.

The first Texel sheep came to Great Britain in the 1970s. It passes on excellent muscle development and leanness to its progeny out of crossbred ewes. It is a very quick-growing hardy breed.

I have kept a Texel ram and put him on Texel cross ewes producing first-class finished meat for the butcher, gaining high prices at market. I have also used a Texel ram on Kerry Hill ewes, again producing very good lambs – not quite as fast to grow as the Texel × Texel crosses but quicker-growing than pure-bred Kerry Hill lambs. Top Texels can command a high price, and much semen is sold. In 2009, at the Scottish National Sale at Lanark, one ram lamb topped the breed price of 220,000 guineas (£231,000).

Other continental terminal sire breeds include Beltex, Berrichon du Cher, Charollais, and the Vendeen.

Texel in-lamb ewes.

Crossing Group

This group consists of sheep that are used to cross onto other breeds to improve the lambs in some way – often to get more meaty lambs and quicker-growing lambs.

Bluefaced Leicester

This breed originated in the late 1800s and early 1900s near Hexham in Northumberland. They have a Roman nose with no wool on the face, and ears that are erect. The wool is tightly purled and looks like small knots. It is of high quality with many uses including hosiery and can be mixed with mohair and silk.

Bluefaced Leicesters – these three shearling rams are the crossing type.

There are two types of Bluefaced Leicester – the traditional and the crossing type. The traditional type has head skin that is blue, which shows through the white hair on the face; the legs are white with no wool. This type is good to use on white-faced ewes such as Welsh Mountain and Cheviots; their lambs are classified as white-faced mules.

The crossing type often have darker markings on their faces and legs. They have been developed to breed dark-faced mule ewes. These rams can be used on many types of hill ewes such as Beulah Speckled Face, Scottish Blackface and Swaledale.

Kerry Hill

This breed originated in Kerry, a small town in Powys on the English–Welsh border area, and is not Irish as the name suggests. The Kerry Hill Flock Book Society was founded in 1899. They have a white face with a black patch around each eye and a black nose; an odd black spot on the face is acceptable. The legs are black and white, quite often white with black knees. Their ears are set high and are erect. They produce a good dense fleece that is free from hemp and is one of the softest of British wools. Rams are used to cross with hill type ewes to produce better conformation than the pure-bred hill breed. Kerry Hill ewes can be crossed with terminal sires such as Suffolks. These lambs are very good, mature early, and have good conformation and length and quality lean meat.

They are excellent mothers, and the lambs are soon up and drinking. I started my flock in 1979 with two ewe lambs and have kept Kerry Hill sheep ever since. As many as

Kerry Hill sheep at Oak Tree Farm – these were all home-bred and exported to Holland.

possible (both rams and ewes) are sold as pedigrees including selling a homebred ram to Kerry in Powys and exporting sheep to Holland, but those that are not good enough to breed from are sold as lambs for the freezer or lambs at market (they may not be good enough to breed from, as they may have a fault such as a lot of black wool in the neck). The lamb is full of flavour, and each year we have at least one in our freezer.

They do well at agricultural shows because of their striking appearance and can win interbreed championships. I find it wonderful to look out of our windows and see Kerry Hill sheep grazing just outside the bungalow, with their well-defined black and white markings and upright ears – the pleasure I get from looking at them is better than a lot of things.

Other crossing breeds include the Bleu du Maine, Border Leicester, Clun Forest, Devon Closewool, Dorset Horn, Polled Dorset, Lleyn, Rouge, Wiltshire Horn and Zwartbles.

Wool Group

These sheep like good conditions, and they produce a lot of wool. They are not as common as they used to be.

Cotswold

It is thought that the breed originated on large Roman estates around the Cirencester area. The Cotswold Hills were the centre of the wool trade in England, and by the

Cotswold ewe.

Middle Ages there were thousands of Cotswold sheep. The sheep played a big part in the development of towns and villages in the Cotswolds.

They were a very popular sheep indeed but sadly became rare after the First World War. By the 1950s they became very rare, but I am pleased to say the number of flocks has since increased. However, the Rare Breed Survival Trust have got them on their 'at risk list.'

They are a very tall sheep. They usually have a white face and legs, and are polled. They have a forelock, which in my opinion makes them look very attractive. They produce excellent wool (known as the Golden Fleece), which is very popular with spinners and knitters. This could be another enterprise for the first-time farmer – producing knitted goods from their wool. They are a hardy breed, producing good meat that is mild in flavour.

Wensleydale

This breed can be traced back to a ram, called Bluecap, born in 1839 in a hamlet near Bedale in North Yorkshire. A very large native breed, it has a deep blue face and ears. The breed produces the finest-lustre wool in the world, the fleece is very long and curly, and the sheep have a forelock of wool, which is known as the topping. The wool is used for the highest-quality knitted garments.

Wensleydale sheep.

Both rams and ewes are polled. As well as producing this excellent wool, it is a crossing sire: many are bred to cross on to hill ewes such as Scottish Blackface. The breed is very scrapie-resistant, producing progeny that are also very resistant. The ewes are prolific, producing two or even three lambs. Black Wensleydales also occur, and a separate register is kept in the flock book.

Other woolled breeds include the Devon and Cornwall Longwool, the Leicester Longwool and the Lincoln Longwool.

Mountain and Hill Group

The mountain sheep live in the highest and hardest land, where it is wet, cold and windy in the winter. The sheep must be very hardy to survive these conditions. The sheep are not very prolific, but they are good mothers, and the lambs are soon up and drinking, and want to survive. Some sheep are more suited to the grass hills. These are also a hardy sheep, but not as tough as the mountain breeds and are more suited to the hills, which have less harsh conditions. They do well in the lowlands as well.

Derbyshire Gritstone

This is a really outstanding British breed that originated in the Peak District of Derbyshire and is bred up to about 2000 feet above sea level. It is found mainly in Lancashire and the Peak District, but is found in other parts of the country as well, but not on the top of the Scottish mountains.

Derbyshire Gritstone in-lamb ewes at Oak Tree Farm.

They do well in lowland conditions. I could see the potential of this breed, sensing I could breed good sheep here in the lowlands and then sell the pure-bred rams and ewes at the breed society show and sale at Clitheroe in Lancashire, or sell them privately, with the surplus sold as good-quality finished lambs for the butcher, and the bonus for me, a breed good to show.

There is no wool on the face and legs, and they have well-defined black and white markings, polled, with ears that point down. The breed produces good wool, which is used for hosiery and knitted garments, and produces good lambs with good flavour.

Herdwick

This breed is native to the central and western Lake District. The name Herdwick derives from the old Norse 'herdrych,' meaning 'sheep pasture.' They are very hardy indeed, grazing up to 3000 feet, and on many occasions they have been found alive under snow drifts.

Herdwick.

The rams have horns. They are white-faced with a blue-grey fleece. The lambs are born black with white ears, but as they get older they turn brown and then blue grey, and their faces turn white. The wool is kempy and coarse, and is used for carpet making and insulation, and is also used undyed for knitwear. It has a dark, well-flavoured meat. They are left on unfenced areas and do not wander off their traditional pasture; this is 'hefted to the fell.' Lake District farms are sold with the sheep, and so new farmers inherit the sheep with the farm. They are mainly found in the Lake District, but flocks are found elsewhere. Small holders may keep a few, and I see some at agricultural shows in this area.

Other mountain and hill breeds include Badger Face Welsh Mountain, Black Welsh Mountain, Exmoor Horn, Jacob, Norfolk Horn, North Country Cheviot, Shetland and Swaledale.

Easy Care sheep

This is really a lowland grazing sheep. It is a relatively new breed of sheep, developed in the 1960s. The breed combines characteristics of the Wiltshire Horn and the Welsh Mountain. As the name suggests, they are easy to care for. They shed their wool in the summer and so do not need shearing. They have very few problems with fly strike and require minimal veterinary care. They will lamb outside, lambing easily with a high lambing percentage and quick growth of the lambs. They are a low-cost sheep with good returns – an up-and-coming breed.

Easy Care ewes and lambs.

Some of the sheep in the main four groups (terminal sires, crossing group, wool group and mountain and hill group) are rare breeds. There are other rare breeds and if you want to specialise in a rare sheep breed you must choose your breed. You may become an enthusiast, show them and sit on the breed society committees, and even become a judge. The Rare Breeds Survival Trust publishes a list of all rare breed animals. Here is a watch list for sheep.

Categories	Registered breeding females	Breed
Critical	Less than 300	Boeray
Endangered	N/A	
Vulnerable	500-900	Castlemilk Moorit, Devon and Cornwall Longwool, Leicester Longwool, North Ronaldsay, Teeswater, Whitefaced Woodland.
At risk	900-1500	Balwen, Cotswold, Hill Radnor, Lincoln Longwool, Manx Loaghtan, Norfolk Horn, Oxford Down, Portland, Soay, Wensleydale, Whiteface Dartmoor.
Minority	1500-3000	Border Leicester, Devon Closewool, Dorset Down, Dorset Horn, Greyface Dartmoor.

That's the end of this chapter, so all's wool that ends wool – to be contin-ewed.

Chapter 13

Sheep Diseases and Ailments

Abortion

Ewes that have aborted should be isolated, and aborted lambs and material destroyed. If abortions are to be investigated, however, it is very important to keep placental tissue as well as aborted lambs.

Enzootic (Chlamydial Abortion)

The placenta becomes infected with a virus; this is the main cause of abortion. It is transmitted from aborting ewes to other ewes and occurs during the last 4 weeks of pregnancy. Sometimes lambs are born alive but die soon after birth. Ewes rarely abort again but continue to spread disease at each lambing. A vaccine is available.

Toxoplasmosis

This is caused by a protozoan parasite, and it can cause coccidiosis in cats. Neutered adult cats are low risk, but cats having kittens in the forage stores create high risks. The parasite produces oocysts (eggs), which are expelled out in the cat's faeces. They can then be ingested by the sheep when eating or drinking contaminated food and water. The food could include grass, hay or concentrates. The oocysts find their way to the placenta and foetuses, which are aborted. The ewe aborts or weak lambs can be born but usually die. Infected ewes do not abort in their second pregnancy – so there is no need to sell them. If it invades in the first 2 months of pregnancy, the foetus will be reabsorbed, and the ewe will return to the ram. This is a common reason for high levels of barren ewes.

Infection in mid-term will cause death of the foetus and abortion about 10 days later or mummification about 40 days after infection.

If you get an abortion in your flock, you must get laboratory tests done to see which infection you have got. Vaccines are available – a live vaccine is used.

Pregnant women should keep out of the lambing shed and well away from new born lambs and afterbirth. They are at risk both for themselves and for the unborn child; these risks are low, however.

Chlamydiosis can cause still birth or abortion in humans. Toxoplasmosis can also cause still birth, miscarriage or birth defects. Listeriosis can also cause miscarriage or premature birth.

Q fever is an uncommon disease in Britain with no recent cases but can cause prematurity, low birth weight or miscarriage. As well as keeping out of the lambing area mothers to be should also avoid handling dirty clothes that have been worn by partners or farm staff. Partners should always wear protective clothing and wash well after being in contact with in-lamb ewes. Never use a hairdryer to dry and warm lambs, as this can create an aerosol of infection.

Acidosis

The overfeeding of concentrates produces a high level of acid in the rumen, and so the rumen and blood pH drops rapidly. The sheep has abdominal pain, looks depressed and can die.

Drench with baking soda, which will reduce the acid, and inject with intravenous concentrated vitamin B. Cure dehydration with glucose saline solution. When starting to feed concentrates to ewes or lambs, start them on just a small handful and increase the ration slowly.

Bloat

The rumen fills with trapped gas, causing it to distend, but it is less severe in sheep than in cattle.

Frothy Bloat

Sheep grazing lush wet pasture are susceptible. The fermenting food in the rumen forms a foam and the gas it contains can not be expelled. A defoaming agent is used to get rid of the foam and is given by a tube into the rumen.

Frothy bloat can happen when the sheep has been eating rich grass in spring or autumn, particularly if it has a high clover content or when eating lots of rich concentrates. The animals can soon die. It can affect a high number of animals.

Mild cases may be treated with cooking oil, but it is best to consult your vet. Obviously if you get an outbreak of frothy bloat, you get them off the grass as soon as possible and give them straw or hay to eat.

Free Gas Bloat

Gas is trapped in the rumen because something is obstructing the oesophagus. This could be a root crop such as a swede. A gastric tube can be used to push the object down, or a trocar can be used to puncture a hole in the rumen, but only as a last option.

Blowfly Strike

When the warm and humid weather comes in May and June onwards, you must watch out for blowfly strike. You will soon become experienced at spotting a sheep with blowfly strike. It is caused by greenbottle flies, which are a complete nuisance. They are attracted by wool that is covered in faeces or urine. The female can lay 200 eggs in one go every day throughout its life. Maggots hatch and feed on the skin and flesh of the sheep. The wool becomes wet, matted and discoloured, and the smell attracts even more flies, which in turn lay more eggs. This causes the sheep much stress and discomfort. They will rub themselves, wag their tails constantly, keep looking round at their tails, bite the infested area and stamp their feet. If left untreated, they may die. When maggots are found, you trim off the wool in the area where the maggots are situated, remove all the maggots and pour on a fly strike product obtainable from your veterinary surgeon or from stores such as 'Countrywide.' If the attack is very bad, an antibiotic injection is also advisable. However, it is best to prevent fly strike rather than treat it. Pour-on products are available containing insect-growth regulators – you will have to use this product at least once a year and perhaps twice. In warm weather, it is best to shear your sheep as soon as possible. You will get fewer blowfly attacks after shearing. Tail docking the lambs in their first week of life means there is less area for the flies to lay their eggs. Most commercial sheep have their tails docked. However, traditionally, many pedigree sheep breeds are left with longer tails, which looks good but is not such a practical option – Derbyshire Gritstone rams have no part of their tails docked.

Bluetongue

Bluetongue is a notifiable disease that affects cattle, sheep and goats. It is caused by a virus spread by certain types of biting midges. It cannot be transmitted directly between animals. It is characterised by changes to the mucous linings of the mouth and nose, and the coronary band of the foot. Vaccinations became available using inactivated vaccines, but currently these vaccines are not allowed. On 5 July 2011, Great Britain was declared free of bluetongue.

Clostridial Diseases

Clostridial bacteria multiply very quickly indeed and produce toxins that are absorbed and cause symptoms; different toxins will cause different diseases. A vaccine that treats all the clostridial diseases and pneumonia is available, and sheep must be vaccinated, in my opinion. A popular vaccine is Heptavac P.

Blackleg

Clostridial bacteria live in the soil and are ingested by the sheep; the bacteria then enter the muscle cells, usually the leg, and remain dormant until the animal has been injured and damaged, possibly through shearing or fighting. The bacteria reproduce very quickly, and produce gas and toxins. The animal may show symptoms of walking with difficulty and have a high temperature, and may die in a few hours.

Braxy

This is seen in older lambs and ewes that have not been vaccinated during autumn and winter. It is associated with ingestion of root crops, and eating frosted, green food will cause indigestion, which may induce it. The first sign will probably be a dead lamb. There is no treatment.

Lamb Dysentery

This attacks lambs 2–5 days old. The bacteria is found in the soil and can be ingested by lambs when sucking from their mother. The intestines are infected, causing diarrhoea, which is often blood-stained, and then enteritis may set in, leading to death.

There is no treatment for those that have it, but unaffected lambs should be injected with lamb dysentery serum as soon as possible after birth. This is expensive, and vaccination of the ewes is cheaper and more effective in the long run.

Malignant Oedema or Gas Gangrene (Big Head Disease in Rams)

This is caused by a soil-borne bacteria that infects the animal through a deep wound. The wound swells up and discharges, and the animal can die after just 1 or 2 days. The rams may have started this off through fighting. Antibiotics such as penicillin, can save the animal if caught early enough; usually, however, it is too late.

Pulpy Kidney

Pulpy kidney or enterotoxaemia is another clostridial disease. It affects lambs usually over 1 month old, but it can affect older animals. Lambs that are doing very well and grazing lush grass or lambs that are given a sudden increase in concentrates are susceptible to the disease. They are usually found dead and show no symptoms. I have seen only one lamb showing this disease that twitched and had convulsions, and was dead within an hour. However, I have found dead lambs that, after a postmortem, had definitely died of pulpy kidney. At postmortem, the kidneys look soft and pale or bloody. Basically, toxins are produced by bacteria in the intestines, and these toxins are absorbed into the bloodstream. Blood vessels are damaged throughout the body and also in the brain – sadly the lamb dies. You can't treat pulpy kidney because the lamb dies very quickly, so make sure your sheep are vaccinated against it with a product such as Heptavac P.

Tetanus

Thankfully, this is quite rare in sheep. The bacterial spores are found in the soil and especially manure. Lambs are affected through dirty cuts and wounds, such as shearing cuts and bites from dogs. It is mainly a disease of lambs, but adult sheep can be affected.

Lambs can be infected, but symptoms may not show for 2 weeks after infection. They find it difficult to chew and swallow, the jaws become locked, the animal becomes very stiff and walks badly, and the muscles become rigid. Death will probably occur 7–10 days after the symptoms, but animals sometimes recover. Mild cases can be treated with penicillin.

Copper Toxicity

Sheep accumulate copper in the liver more readily than cattle and pigs, and it is the accumulation of copper in the liver that results in copper poisoning. I have experienced this when I first started showing Kerry Hill sheep in the 1980s. To get big strong sheep, they were encouraged to eat as much concentrates as possible, which was a coarse mix for sheep and cattle. I did not realise at the time it had got too much copper in the feed. Two shearling ewes were off their feed, and their skin and eyes turned yellow with jaundice. Their urine was probably red brown because it would contain haemoglobin, but I did not see that. Pig feed also may have very high copper content and must never be fed to sheep.

Treatment is to give a drench of ammonium molybdate and sodium sulfate or prussiate of potassium, but both my ewes died, which was to be expected.

Entropion

This is when the eyelid rolls inwards, usually the lower lid. The eyelashes on the eyelid brush against the eye and cause discomfort to the eye, and so the eye runs, and stains appear under the eye. If untreated, the eye may become ulcerated and it can even cause blindness. An injection into the rolled eyelid or a clip in the skin is the best treatment.

I bought a Texel ram that sired some of his lambs with entropion; the condition is inherited, so I did not want to use him the next year because of this problem. It was not fair to sell him on as a breeding ram, so he went for slaughter.

In very mild cases, I have pulled on the skin with my finger and thumb at the bottom of the eye, three or four times a day for a few days. This has corrected it on a few occasions, but the injection or skin clip treatment is certainly the best.

External Parasites

Lice

Lice can cause the fleece to go ragged, and wool is lost. Sheep will rub against shed walls and fencing posts, and bite their own fleece, because the lice irritate them as they feed on the skin and wool. Surprisingly, they are found mainly in the winter. They can live away from the sheep for 17 days in buildings, on fence posts and in other suitable places.Insecticide treatments are available, including spot-on and pour-on products. A good, popular pour-on product is Dysect.

Sheep Scab

Sheep scab mites feed off the skin and wool, causing the sheep to be very uncomfortable and causing lesions and scabs, which get bigger and spread. Affected sheep will rub against fencing posts and buildings, bite their wool and dig with their back legs. Transmission can happen when sheep are brought in tight to one another in pens. The sheep look awful, they lose condition, and the wool is lost. It can be confused with an attack of lice. It can also be spread indirectly at markets. Control using diazinon-based dip or ivermectin injections or similar products.

Ticks

Ticks are found in rough pastures, hillsides and moorland. They irritate the skin by feeding off the sheep, and they can carry disease such as louping ill. Hefted flocks build up an immunity, and bought-in stock from lowlands will be much more susceptible. Ticks are found in two seasons, spring and autumn. Pour-on insecticides such as Dysect are good and provide control of ticks for 8–12 weeks.

Foot and Mouth

This is a highly contagious virus indeed and is the stockman's complete nightmare. It affects cattle, sheep, goats, deer and pigs and can be carried in many different ways including car tyres, manure, wild birds and winds.

Sheep have blisters on the tongue, mouth and teats, and between the hooves. The sheep become lame and listless. Infected and in-contact animals are slaughtered, and all carcasses buried or burned. When it was in this area in 2001, I was lucky that my sheep did not get it. An old wives' tale is that if you hang raw onions up in your sheds, you will not get it. I hung up the onions and did not get it, but I am sure it was not due to the onions.

Foot Rot

If a sheep appears lame, there is a good chance that it has foot rot or scauld. Foot rot is a contagious disease caused by bacteria. The bacteria thrive in warm, moist conditions and actually rot the foot away. The area between the toes becomes infected, and the sole of the foot is attacked, causing separation of the horny tissue. Once you have smelt it, you will not forget it because it really does smell foul. The bacteria need infected animals to live on and do not survive for more than 2 months in the soil, but they will survive for long periods in infected hoof clippings.

Sheep with foot rot should not be foot-trimmed unless severely overgrown, and sheep should not be foot-trimmed routinely unless the foot is visibly overgrown/deformed. The best treatment is by injection of long-acting antibiotic (oxytetracycline) and spray. You should get 75% cure from one injection if given promptly, and a second dose will clear up the majority of what's left. Any hoof clippings should be collected and burned.

Zinc sulphate in a foot bath is useful to stop the sheep getting foot rot, or it can help to cure it. In lambs, I get a few with abscesses between the toes which causes lameness, and the adult rams fight and become lame – they usually injure a front leg or a shoulder.

Hypothermia in Lambs

If found to be very cold, lambs should be brought into the warm and given an intraperitoneal injection of 20% dextrose (glucose) solution. If a lamb is wet, it should be towel-dried, then warmed up using a warming box or heat lamp, and then fed colostrum. Return the lamb to the lambing shed when the lamb is up and appearing normal.

Hypomagnesaemia (Grass Tetany/Grass Staggers)

This is a metabolic disorder and is due to magnesium deficiency. Low levels of magnesium or mineral imbalance in the grass, i.e. too much nitrogen and phosphorus applied to the grass, may lead to magnesium deficiency. The sheep will show nervousness with twitching muscles, will stagger and will be off their feed, with convulsions and possibly death. Treat by giving magnesium injections. The sheep may also be low in calcium, so it is usual to inject both magnesium and calcium.

Joint Ill or Navel Ill

This is caused by bacteria. Dirty sheds and lambing pens will encourage this disease. Lambs are affected up to about 4 weeks old. Leg joints become swollen, painful and hot, and the lambs can have a fever. If it is caught very early on, daily injections of penicillin and streptomycin, or broad-spectrum antibiotics such as tetracycline may work, but some lambs will probably die. If they live, they often remain lame and do not do as well. Prevent it by smothering the navel at birth with iodine and make sure the lambs drink colostrum. I always dip lambs' navels with iodine, but I have still had the odd case of joint ill. Most lambs that get joint ill or navel ill have not had sufficient colostrum.

Listeriosis

This is a bacterial disease. It does not spread by direct contact but spreads by the sheep ingesting it. Mouldy silage can contain the bacteria, and cuts and abrasions in the mouth can be a path of entry into the body. Punctured big bale silage is a very high risk.

Infected sheep will be disorientated and walk in circles. The sheep often die. I will not feed silage to sheep. Antibiotics may treat the disease if given in high doses – penicillin and streptomycin may cure it.

Louping Ill

This is an acute viral disease that affects the brain and is transmitted by ticks. It is found in April and June, and again in September and October, and mainly affects sheep, but cattle, pigs, goats, red deer, horses and humans can all get louping ill. It is also a significant disease of grouse.

The sheep will get tremors of the neck and be unsteady; convulsions follow and eventually paralysis and death occur.

There is no treatment to control the disease, it is best to prevent the ticks attacking the sheep; use a pour-on insecticide such as Dysect. A vaccine is available, which is given as one single subcutaneous injection.

Liver Fluke

Sheep of all ages can be affected. The liver flukes live in the liver and bile ducts of the sheep and will lay eggs, which pass out in the faeces. They hatch into a form called miracidia and then pass into a mud snail and emerge back onto the grass. They are then called cercaria. They move on to the grass turning into metacercaria and are eaten by sheep. Inside the sheep, they move to the liver and bile ducts, and can cause death.

Affected sheep will suffer from anaemia; gums will be especially pale; they may lose weight, scour and have a swollen belly; and they may have swellings under the jaw. They will suffer liver damage. The parasites can be brought on to the farm from new sheep. Sheep should be dosed with flukicide. Draining the land will help get rid of the wet environment for the intermediate host, the snail.

Maedi Visna (MV)

This is a chronic disease caused by a retrovirus. It was brought into Great Britain from sheep imported from the continent. Both the lungs and the central nervous system are affected. It can show up as pneumonia, paralysis, mastitis or arthritis, and animals will lose weight. All sheep can be infected, especially lambs. It is spread by direct contact with the ewe's milk, which transfers the virus to lambs. It has a long incubation period.

There is no treatment. If it is found, the progeny in the flock should be sold for slaughter to stop it spreading to other farms. There is an MV Monitoring Scheme and an Accreditation scheme. MV-accredited sheep undergo tests and, if free of the disease, must be kept separate from non-accredited sheep – this can be seen at agricultural shows.

Mastitis

This is a bacterial disease that occurs a few days after lambing, or later on in the production of milk. The bacteria penetrate the teat canal and get into the udder. It can be broadly divided into these forms:

Subclinical Mastitis

This is a low-grade mastitis with no visible changes, but this can still result in a 'blind' quarter at next lambing.

Mastitis

The ewes produce less milk and develop clots in the milk, and lumps form in the udder. It may go unnoticed until the udder is examined and seen to be infected. Broad-spectrum antibiotics, such as a long-acting tetracycline, are effective and can be injected. Intramammary tubes of antibiotics are placed up the teats. Any ewes with hard lumps in their udders (scar tissue) are best culled, as they will not produce enough milk, and the affected side may have a 'blind' teat at next lambing. Here at Oak Tree Farm, we are careful to wean the lambs correctly. When the lambs are taken away, the ewes are put on a diet of straw or poor grass to stop milk production and so prevent mastitis.

Severe Mastitis

The ewes will not want to walk; they may hold a foot up in the air and refuse to let their lambs drink. If left, it turns nasty, and the udder turns black. Gangrene sets in, and the affected side of the udder may drop off. A ewe could recover following antibiotic treatment, but she will only have one side (quarter) to the udder next year, or indeed both sides could have been infected.

Milk Fever or Hypocalcaemia

This is a metabolic disease caused by calcium deficiency, which affects in-lamb ewes, mainly in good condition, which are close to lambing. The ewes will stagger and then go down; if untreated, this will progress to a coma and possibly death.

The disease is easily treated by calcium borogluconate injections subcutaneously, or in serious cases the calcium should be given by slow intravenous injection.

Orf

This is a virus disease, usually affecting lambs. Pustules and ulceration appear on the lips and mouth. Orf can also affect the genitalia, udder and feet. You must be careful when handling sheep with orf because humans can also contract the disease. Lambs usually recover, but sometimes lambs have died, as they have not been able to suck milk. The infection lasts 4–6 weeks. The virus can survive for years in the lambing pens and buildings but cannot survive outside in the winter. The lambs will get better with no treatment, but aerosols containing tetracycline are effective to treat the secondary infection. When you see it, you should vaccinate the rest of the flock. Do not vaccinate lambs that have got it because you are giving them more of the virus.

If you have got a problem year after year, you are best to vaccinate, but if you have not seen it for a long time, you can hopefully get away without vaccination. Orf has been present on this farm, and I have vaccinated, but I have not seen it for about 8 years, so we have ceased vaccinating. If you have never had it, do not start to vaccinate.

Pink Eye

This can be caused by different infective agents including mycoplasma (bacteria). The eyes become cloudy and then pink. You will notice tear staining under the eye. Antibiotic cream placed on the surface of the eye works well. Sheep on this farm have suffered with pink eye in the past, but fortunately not for a few years. It tends to be spread by contact, and trough feeding can result in high levels of spread once the disease is present.

Pneumonia

Pneumonia is an inflammation of the lungs caused by microorganisms. Pneumonia is found in lambs and adult sheep. One of the most important forms is Pasteurellosis. Pasteurellosis is found usually in adult sheep, but it can cause septicaemia in lambs 4–6 weeks old. Sheep lose their appetite and breathe rapidly, have a high temperature and may have a dry cough. A postmortem will confirm the disease. Pneumonia is susceptible to antibiotics. Most sheep farmers I know use a vaccine that is mixed with the clostidrial vaccine. Heptavac P is used here on this farm.

Pregnancy Toxaemia (Twin Lamb Disease)

Pregnancy toxaemia, ketosis or twin lamb disease can occur in ewes during the last weeks of pregnancy. It affects older, thinner ewes and sometimes fat ewes carrying a large foetus, twins or triplets. Ewes will stop eating and may stand with their heads down or up. They sometimes go blind, grind their teeth and refuse to get up, and if you don't treat them they will die. The cause is low blood sugar (glucose). The uterus will have enlarged, and there is less room for food in the rumen. It is therefore essential to feed a concentrate ration with high energy to avoid this problem. Treat with propylene glycol – 200 ml three times a day with 3–4 litres of an electrolyte solution to stop dehydration. The disease can be triggered by stress. I feed my sheep well during the last 6–8 weeks of pregnancy but I must admit I have had the odd case of pregnancy toxaemia. Feed well, but if you do get it in a ewe, hopefully you will have spotted the signs early on, and you will save your ewe. If it occurs within 2 weeks of lambing, it may be best to induce birth with corticosteroids.

Ringwomb

In this condition, when the ewe is about to lamb, the cervix does not open up properly to allow the lamb into the vagina. I have managed to open them up in the past with gentle massage to the entrance to the cervix, but more than likely a caesariean has to be done to get the lambs out. One was completed here in 2012, our vet delivering two live lambs, and the ewe reared them. However, it was on the coldest night in winter, with me holding the ewe and our vet working hard.

Schmallenberg

Schmallenberg virus is a new disease that was first detected in 2011 in Germany. It has now spread into Britain and in fact during 2013, tens of thousands of lambs have been infected. It is not a notifiable disease, but affected farms have been requested to report cases to their veterinary surgeon. It can cause a mild illness in stock, but the main problem is that it causes congenital malformations and still births in cattle, sheep and goats. Lambs are born deformed in shape but do not have bits missing or extra body parts. It is transmitted by midge bites. In sheep, it appears that the early lambing flocks are mainly affected, because they will have been put into lamb during late summer or early autumn when the midges are most active in the warm weather. The later-lambing ewes may have had more time to develop a natural immunity to the virus. There is no treatment at the moment, but a vaccine is now available.

Scours or Diarrhoea

Scours is really a symptom of another complaint or disease such as lamb dysentery. Adult sheep and lambs can suffer from scours, and a scouring sheep will attract blowflies and so possibly get maggots.

Infectious causes of scours are bacteria, viruses protozoa and worms. Non-infectious causes are problems with feed, pasture and water, poor management and stress.

Antibiotics are a good treatment for bacterial infections but not for viral. Isolation and oral electrolytes will avoid dehydration, and this is about the best you can do. Coccidiosis is best treated with an anticoccidial agent orally. Faeces samples will need to be looked at in the lab to find out what is causing the scours, in order to apply the best treatment.

Scrapie

This disease is caused by a prion, a small infectious particle composed of abnormally formed folded protein that causes progressive neurodegenerative conditions. The sheep will scrape their wool against fence posts and buildings, and any other suitable place because of skin irritation, and then wool loss occurs. The head is carried high, with a high stepping gait of the front legs, and convulsions occur. Death occurs after up to 6 months of the disease.

No treatment is available. Some breeds are more susceptible than others. The National Scrapie Plan, which encouraged farmers to breed from sheep that were more resistant to the disease, is no longer in existence in this country but it has been hugely successful in reducing the disease. Try to buy rams that are fully scrapie-resistant.

Swayback

This is copper deficiency. Copper is essential for life, but too much copper leads to copper poisoning. Lambs suffering from swayback (milk is not a good source of copper) are affected at birth or when they are 2–6 weeks old. The infected lambs show weakness in the hind quarters and sometimes cannot stand or walk.

To treat, give a safe copper product orally, as recommended by your vet. You must dose carefully, because copper poisoning could result. It is much more effective to supplement pregnant ewes than trying to treat affected lambs.

Urinary Calculi

This is a metabolic disease of rams and castrated rams. The urethra becomes blocked by calculi (stones or crystals). The ram finds it difficult to urinate, as the urine may just dribble out or does not flow at all. The ram will be in pain, with a distension of the bladder and urethra, and may stand with a humped appearance. I have seen it with a few of my Kerry Hill rams that have been pushed on a diet of concentrates, to get them into show condition, and I have seen it at shows, when a ram starts off with this complaint. A phosphorus/calcium/magnesium imbalance or excess in the ration would contribute to the problem. Rams need to drink plenty of clean water when on a high concentrate ration.

I have drenched rams with lemon juice, so that the acid will hopefully dissolve the calculi. If this does not work, a vet will need to cut off the feeler at the end of the penis. Never feed growing lambs on ewe nuts, as the high mineral content often results in this problem.

Uterine Prolapse

This condition occurs after lambing. As the ewe keeps straining, the uterus or womb becomes turned inside out and protrudes out of the vagina. The prolapse has to be pushed back in, and this is best done by a vet; it will certainly want cleaning before it goes back in.

It can be caused by calcium deficiency, retained afterbirth, or inversion of the tip of the uterine horn. Afterwards, injections of antibiotics and anti-inflammatory pain killers will be beneficial.

Vaginal Prolapse

This is when the vagina comes out of the body and hangs like a big ball under the vulva. It is found during the last month of pregnancy or soon after lambing. It could be due to many causes, including being over-fat or too thin, not enough exercise and too much fibre in the diet.

I have experienced a few of these over the years. I wash it with warm soapy water and mild disinfectant, and then push it all back in. A plastic retainer 'spoon' can then be inserted and fixed to the sheep to hold it in place. I have used these with only mixed results; they did not always work for me, as the sheep would prolapse again. The best way is to use a suture made of nylon tape, which needs to be removed just before she starts to lamb.

I sell the ewe after she has reared her lambs, as I do not want the bother of it occurring again – as it probably would. It can be hereditary.

Watery Mouth

This is a bacterial disease in which lambs will scour, have a cold mouth and salivate. Many will die, usually within 12–24 hours after showing symptoms. It is found in dirty conditions, with buildings and lambing pens that want a good clean out. Lambing pens and buildings need to be kept clean. The lambs need rehydrating and treating with antibiotics. I have seen watery mouth on a neighbouring farm but, I am pleased to say, never on this farm.

Worms

Parasitic worms can cause death in sheep if the symptoms are not spotted. These worms live in the stomach, intestines and lungs. Loss of flesh and scouring (diarrhoea) are the main symptoms. Sheep grazing poor pasture and those that are overstocked are susceptible, and lambs may show symptoms when they have been grazing for just a few weeks. Worm (drench) your sheep and change your wormer frequently; sadly worms are proving resistant to many of these products.

Other Conditions

There are many more diseases and complaints associated with sheep. The most unusual I have experienced over the years was three lambs on three different occasions born with no anus. On one of these lambs, the veterinary surgeon made a small hole in the rectum, and the lamb was fine. The other two had no rectums and had to be destroyed.

Mary had a little lamb
Its fleece was white as snow
And everywhere that Mary went
The lamb was sure to go
Now Mary found the price of meat too high
Which really didn't please her
Tonight she is having leg of lamb
The rest is in the freezer.

Remember: prevention of disease is better than cure.

Vaccination for clostridial diseases.

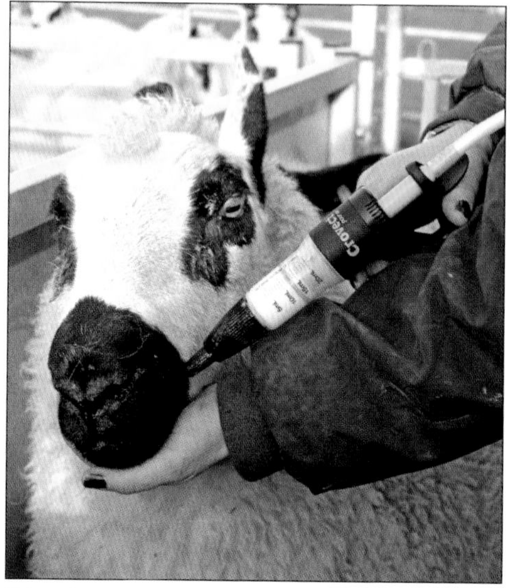

Worming.

Chapter 14

Making a Start with Pigs

My first pig was called Ink because he kept running out of his pen! Pigs are kept for their meat (approximately 9 million pigs are slaughtered each year in Great Britain), some are sold as breeding stock, and recently some have been kept as pets, such as the Vietnamese pot-bellied pig and micro pigs, which, as their name suggests, are very small indeed. A pork pig is killed at pork weight – about 60 kg live weight. A baconer is heavier – about 80–100 kg. A cutter is between pork and bacon, and is kept to produce large joints. You can keep pigs intensively for all of their lives, or you can choose outdoor pig keeping, or a combination of the two systems.

Indoor Pig Keeping

My top tip for starting with pigs and starting at the very bottom of the farming ladder is to go out and buy some weaners. These will be 28 days or older. To rear them up to pork weight is easy. You need a building to put them in, some feed and water, straw and your labour to look after them. If you are just renting your field you may not get permission from the land owner to build a small shed; however, if you have bought your field, you can easily construct a small shed to keep two to six weaners in (or more). I have kept weaners in an old bicycle shed. The shed needs to be well ventilated but not draughty. The floor and roof are best insulated, and a concrete floor is needed. I have used a paving slab floor, but concrete is easier to clean, and slabs have to be large or they will be rooted up. It is most important to avoid draughts and cold. The shed can be made of strong timber, but pigs love to force wooden planks off with their snouts. I lined the bicycle shed with metal roofing sheets; there is no need to go up to the roof with them – just line the bottom half. I progressed to buildings made of breeze blocks, which were much better and rendered on the inside for easy cleaning. The roof should have a metal sheet underneath, then some insulation material and then a top metal sheet. If you just have a single metal sheet, you will get condensation dripping on your pigs, which will make them and their bedding wet.

Weaner pigs 4 weeks old.

You can easily keep a couple of weaners in a building about 3.6 m × 3.6 m (12' × 12'). I have kept four in a breeze-block building 2.43 m × 2.43 m (8' × 8'), which was fine when they were small but too tight when they got older.

If your landlord won't let you build a shed, you could try driving around the farms in your area and see if a farmer will let you rent a building to keep pigs in; you might be able to rent a whole set of buildings enabling you to keep many more pigs and perhaps even a breeding herd. If you are going to keep more than a handful of pigs, you really can't keep them all in converted garden sheds or redundant garages.

If you are not living on the premises where your pigs are kept, you need to bear in mind they must be very secure. Escaped pigs at midnight with no one to see them until the following morning could bring you a lot of trouble; also, they want to be under lock and key because a farm theft is a lot easier when no one is living on the premises. Finally, if your pig building is part of a derelict farm, behind a housing estate in a poor, rundown area, you will have to consider vandalism – including setting fire to your straw, windows being broken and pigs let out.

Pigs are clean animals; if they have enough space, they will not soil their sleeping and eating areas. You will, of course, have to clean them out regularly and get rid of the pig muck. If you have your own field or fields, you can spread it (or some of it) on your land; it is not popular with allotment holders or gardeners. If it is a rented field, your landlord might agree to it being spread, but failing this you will need to find a farmer who will take it. It would be ideal just to tip it on his muck heap (don't forget you have got to get it there – the easiest way would be a second-hand trailer towed by your car). He probably won't pay you for this muck, and if you can't tip it on his muck heap, you may have to spread it on his land.

I have spread tonnes and tonnes of muck by mucking out the pigs and then loading it into a small livestock trailer pulled by the car. The back of the trailer was put down in the field, and then the muck was thrown out with a fork. You can't do this in very wet weather because your car will get stuck in the field. I have been stuck even with a four-wheel-drive car.

If you have lots of pigs you may need to buy a tractor and muck spreader or get contractors to spread it for you. You only need to keep these weaner pigs until they become porkers at about 4 months old. When I was teaching agriculture, I used to keep four pigs per term and so rear 12 in the school year – with no pigs present for most of the major holiday time. I taught for 25 years and kept pigs most years, so I must have reared well over 200 with no deaths.

I never sold any of these pigs to a butcher but found customers to take them as half pigs for their freezers. You will need to find an abattoir that will not only slaughter the pigs but also butcher them. You usually need to give them plenty of notice, and you will collect your pork all cut up into joints and chops, labelled and in plastic bags – it is, of course, a lot less trouble to bring them back than to take them unless a bag splits open, and blood leaks out onto your car seats. This has happened to me a few times!

To look after your porkers, you will need a solid water trough that won't tip over, some straw for bedding, a wheelbarrow, shovel and brush, and pig feed. Pig feed is cheaper if you buy in bulk rather than in bags, but for a very small number of pigs you will not need a large bulk load, and you probably won't have anywhere to store it, so it will have to be in bags.

To keep a few pigs indoors, I would go for Large White/Landrace cross types. You would buy these from a reputable breeder. I did this every time. You could go for a rare breed such as Tamworth or Gloucester Old Spot, but these will take longer to grow, and you need to find customers that are willing to pay a premium for rare-breed pork.

Feeding

When your weaners arrive, you will need to feed them at the same time twice a day, and you will need to look at them to check their health, so if you have a full time job and don't get home until after dark, you will need an electric light. Some farmers feed growing pigs ad lib; others ration the pigs. Pigs should eat up their rations within 20–30 minutes; they may eat more in cold weather to keep warm. Larger breeds will eat more than smaller breeds; however, you feed approximately 450 g (1 lb) of feed each day for each month of age taking it up to a maximum of 2.75 kg (6 lbs) of feed.

Compound feeds are very convenient and are formulated to be a complete ration, and so include everything the pigs need. For larger scale pig farming you will need creep feed, grower and finisher rations plus rations for your sows and boars, and lactating sows. Pig feed can be purchased in various forms such as meal pellets, cubes, rolls and nuts. You will need to discuss with your feed-firm sales representative the best rations for your pigs, taking into account the oil, protein, fibre and ash content. Once you have done this, get a quote from another feed firm to see which 'like-for-like' ration is the cheapest, and it should be 'like for like' and not just a cheap ration with much less protein for example.

It is cheaper to mix your own ration and even cheaper if you can grow your own barley or wheat. I have mixed rations in the past with a shovel on a concrete floor, which is hard work. Pig units with many pigs can justify a mixer wagon, which is powered by a tractor. If you do mix your own ration, you need to buy straights such as soya, barley meal and wheat meal plus minerals. You will need expert advice to formulate your ration. You can grow vegetables for pigs – they especially enjoy carrots and swedes, which can cut down your feed bills. It is actually illegal to feed catering waste or swill that has gone through a kitchen. However, you can feed trimmings from vegetables and fruit as long as it has not gone through a kitchen, so trim your cabbage leaves off outside in the garden. Try to feed your pigs so that you can put feed into the pen without you going in, as they will knock you over in their rush to get at the food. Make sure there is enough trough space, or if feeding on the floor, make sure all the pigs get their fair share. You want your pigs to grow well with no 'poor doers,' but at the same time you don't want your pigs to get too fat. The killing out percentage of pigs is approximately 72–74% – the meat you can use is about 64%. A weighing crate is a valuable piece of equipment, but when you get experience you can look at a pig, and you will have a really good idea of its weight. You can estimate the weight using the following formula (don't forget it is an estimate not an exact weight): you measure the pig's length all the way from the ear to the base of the tail; you then measure the girth behind its front legs or circumference in metres. Then you do a quick calculation – multiply the girth × girth × length and multiply that by 69.3 – your answer is in kilograms e.g. 1.25 m × 1.25 m × 1.04 m = 1.625 m × 69.3 = 112.6 kg. My wife, Sarah, our two children, Jonathan and Roseanna, and I often go out for lunch on a Saturday. Sausage and eggs is on the menu, and I often think to myself that it is just a day's work for the hen but a life-time commitment for the pig.

Another method of getting on the farming ladder is to contract rear pigs. Large firms enter into contracts with farmers to rear pigs for them. The contractor will supply the weaners or growers and will provide the feed. Contracts do vary, but they will probably also provide veterinary care, provide the transport, register for quality assurance and then pay you for looking after the pigs, which could be performance-related with bonuses. You provide the building, and you keep it maintained. You also provide straw, water, gas and electricity plus your labour and machinery. An up-to-date list of the contracting companies can be found with BPEX and the National Pig Association.

If you have got the buildings, this is certainly a good way for first-generation farmers to get involved in the pig industry.

Breeding Pigs

To breed pigs, it is much easier if you are living on site with them. If you have got sows farrowing, it is not really convenient to get in to the car and drive to them; also breeding pigs will take up more of your time because there is much more to do than when just keeping a few weaners.

If you are going to breed pigs, then get some experience on a farm first and try to attend a course at college. You need to determine whether you are going to keep intensive indoor pigs or outdoor pigs or a mixture of the two. You also need to determine what breed you are going to keep, either intensive quick-growing Large White cross Landrace types or a slower-growing rare breed such as the Tamworth, which would hopefully pay you a premium at a farmers' market or at a specialist butcher.

What to Look for in Buying Your Gilts or Sows

I would suggest buying your gilts or sows privately from a reputable breeder rather than at market. If you are unsure, take an experienced farmer with you. Large companies such as ACMC specialise in breeding and selling breeding stock. Young female pigs are called gilts; a gilt will be ready to go to the boar at about 230–240 days old. To choose your gilt or sow for breeding, she needs:

1. to have 14 teats spaced evenly and starting well forward;
2. to have a good-shaped vulva;
3. to be able to walk well and stand up on her toes;
4. to have a good developed rear end and a good length to her body;
5. to not be too fat.

Signs That a Pig Is On Heat

Gilts and sows will be on heat every 21 days but this can range from 18 to 23 days, and this heat period lasts only 1–3 days. You can tell when she is on heat:

1. Her vulva becomes red and inflamed – one of my students in an examination wrote 'she will have a bright red Volvo.'
2. If you press on her back, she will stand still, showing that she is ready to be served.
3. She will have an altered attitude, and she will act differently.
4. She may get a stringy mucous discharge from her vulva.

Watch out for the first heat period, and then if she shows the signs again 21 days later you can get her served.

Boar or Artificial Insemination (AI)?

If you have just the one sow or just a few sows, it is not worth keeping a boar. If there is a large pig-breeding unit near you, they will probably not let you take her there to be served, and they probably won't lend you a boar either in case your gilt or sow introduces disease onto their farm. Some farmers hire out boars – however, they will not be taken off site; you will have to take your gilt or sow to stay on the farm. Your pig or pigs will have to be checked out for any health problems.

AI is a task that you can do. I would suggest seeing it done before you try; there might be a pig farmer in your area that would show you how it is done. Your best bet is to find a small holder with just a few sows, or alternatively if you look on the internet there are farms that run pig courses, or you can contact your nearest small-holders society and ask their secretary if they know of a contact. You can also look for a college course, and the British Pig Association may be able to recommend someone. There are specialist companies that sell fresh or frozen semen, e.g. Deerpark Pedigree Pigs, and they can supply semen from 14 breeds. The semen is stored in straws and will be delivered out to you. The semen is placed into the gilt or sow with an insemination rod, and then you inseminate her again 24 hours later. After insemination, you will watch her carefully 21 days later – if she comes on heat again, she will have to be inseminated again.

Gestation

The gestation period of a pig is the easiest gestation period to remember – 3 months, 3 weeks and 3 days (115 days). Depending on the breed, you can expect 6 to 14 piglets.

Farrowing

The gilt or sow needs to be placed into a farrowing crate 3 or 4 days before she is due. A farrowing crate is necessary to prevent the sow lying on her piglets. You will need an infrared lamp to keep the piglets warm. Signs that she is ready to farrow are as follows:

1. She will try to build a nest out of straw (if she has got straw) within 24 hours of giving birth.
2. Her udder becomes larger, and if you squeeze a teat you should be able to draw out some colostrum.
3. She may start panting, and she will be uncomfortable.
4. A blood-stained fluid may be passed out of the vagina 1–2 hours before giving birth.
5. If pellets that are small and green pass out of the vagina, the birth of the piglets should start within 1 hour.

Sows usually need no help to farrow; once the first piglet arrives, the others soon come followed by the afterbirth. Farrowing will probably take about 2 hours. Piglets are marvellous and will immediately move round and find a teat for a drink.

If the sow is struggling to farrow but looks as if she should be and is showing all the signs, she may need help. Sometimes she may have given birth to one piglet and then there is a long gap of about 40–50 minutes; she will then also need help.

A commercial sow (a mixture of breeds) in farrowing crate with piglets.

Commercial piglets.

You need to clean your hands and one arm. Using lubricant, put your hand into the vagina, and feel around for the problem. Gently pull the piglet out. Once born, you clear the piglet's nose of mucous, make sure it is breathing (if not, slap it), rub it dry with a towel and place it on a teat.

Each piglet has its own teat, the first pigs to be born take the front ones where most milk is produced, and the last to be born have the teats near the back where there is less milk. This first milk is of course the colostrum, which is essential.

On the first day of life, clip the teeth with a pair of tooth clippers – learn how to do this first from an experienced pig keeper. There are eight teeth, and you just round them off. If you leave the teeth, they can cause the sow a lot of pain; the teats will be cut by the teeth, making them sore, which can cause infection and mastitis. If she is sore, she will walk away from her piglets and not let them feed properly, which will restrict their intake of milk. Also on the first day of life, each piglet is injected with iron to prevent anaemia. Piglets are born with limited supplies of iron; anaemia is a deficiency of iron in the blood and can affect piglets during the first 3 weeks of life. Sow's milk is deficient in iron, and so indoors the piglet has no way of getting enough iron until it eats creep feed. In the wild, the piglets would get enough iron from the soil.

Tail docking is also carried out 12–72 hours after birth and not before 8 hours because it is best done after the piglet has drunk colostrum. It is done to prevent tail biting. Pigs will bite each other's tails, which can cause pain and infection, and carcasses can be condemned at slaughter. The tail can be removed with a clean cut with scissors (leave 16 mm), and bleeding stops very quickly. Tails can also be docked using a burdizzo, which crushes the tail and destroys the nerves and blood supply. The tail will drop off after 2 or 3 days. Make sure you have a lesson from an experienced farmer before you carry out this operation.

The piglets would receive creep feed and be weaned at 28 days; they can then be sold as weaners or kept on for pork or bacon. To keep pigs on an intensive system, you could follow the method of some of my near neighbours. Philip and Nigel keep about 180 sows and take the young pigs right up to bacon weight. They use a combination of older buildings and have invested money into new buildings. They make an excellent job of pig production and are keen and interested in what they do; in fact I would call them enthusiasts.

The homebred gilts and sows are a mixture of Large White, Landrace and Meidam crosses. The Meidam is now a popular pig. ACMC is a pig-production company creating high-quality breeding stock. One of the dam lines they have produced is the Meidam, which is exclusive to ACMC. It is white in colour and has a long body and floppy ears. The sows are very prolific, producing a great deal of milk with a good number of teats and a calm temperament, and producing pigs with an excellent commercial carcass.

Philip and Nigel have their own boars, which are Large White with a small amount of Pietrain in them, which are bought in. The Pietrain is a Belgian breed that was not exported to Great Britain until 1964. It is a terminal sire but not a breed I would recommend for the first-generation farmer to breed pure. They are more for the specialist breeder.

The gilts are put to the boar for the first time at 230–240 days old and so will farrow at about 1 year old. They remain in groups in buildings until 3 or 4 days before farrowing, when they are put into the farrowing crates. This herd of pigs averages 12.4 piglets per litter, and the number of piglets reared would be 11.3. During their first day of life, they receive an iron injection, their teeth are rounded off, and the last third of their tails is taken off; no castration takes place. The piglets would receive creep feed pellets from about 14 days onwards.

The piglets are weaned at 28 days old, and at this age they would weigh 7–8 kg. Different litters are then housed together, and the sows go back in with one another and are housed in a building next to a boar. The sows come on heat again between 4 and 6 days later, but usually it is on the fifth day.

Growing pigs 15 weeks old.

The sows would average 2.3 litters a year, but it is possible for them to have 2.5 litters a year. They are fed on pig nuts bought in bulk containing 12–14% protein, but when lactating, the protein content would be up to 18–19%. The sows would eat about 3 kg of feed a day when 'in pig.' Lactating sows would have their rations increased and, at 21 days after farrowing, may be eating up to 10 kg a day. Sows are only allowed to have six litters in their lives because by this time, they would probably be more difficult to get 'in pig,' will give less milk and will produce smaller litters. They would be about 3½ years old and probably wouldn't be wanted by anyone else to breed from, and so they are sold to an abattoir.

The young growing pigs would be fed on a home mix milled and mixed on the farm, which includes wheat plus barley, soya and biscuit meal (waste from a biscuit factory). They are fed ad lib from the day they are weaned until slaughter.

The pigs grow really well and would be moved twice again into larger buildings. They are grouped together in similar sizes, but boars and gilts are not kept separately. Plenty of pigs are kept together in winter months to help them keep warm.

Finished pigs 158 days old.

From weaning to slaughter, they would average a weight gain of 700 g a day, with the best pigs achieving 800 g a day. When they are really growing well, they will put on a kilogram a day.

The pigs are sold privately every fortnight to an abattoir. They are sold at over 88 kg each, averaging 95–100 kg. At this stage, they would have been weaned 130 days ago, plus their 28 days of life before weaning puts them at 158 days old.

During their life, the pigs will undergo vaccinations and medication as follows:

- At 10 days old, piglets are vaccinated against enzootic pneumonia (EP).
- At 21–28 days old piglets are vaccinated against postweaning multisystemic wasting syndrome (PMWS).
- At weaning, the piglets get a booster for EP.
- Medication is added to the feed for *Streptococcus suis* until the pigs weigh 30 kg.
- When the gilts weigh 70 kg, they are vaccinated for erysipelas and then get a booster for erysipelas and parvovirus just before they are served.
- Three weeks before they farrow, they are vaccinated for *E. coli* and scours.

Outdoor Pig Keeping

Keeping pigs outdoors has become more popular for two main reasons: (1) you are not investing a great deal of capital into new buildings, and (2) much of today's public like to think that the pigs are roaming free.

Usually outdoor pig production is where sows are housed outside, and they rear their piglets to weaning. On some farms, some are reared outside, but the vast majority are finished in indoor pig units. These can be units owned by the breeder of the pigs, or the pigs are sold to specialists who take them to pork and bacon weight.

Outdoor pig production. Sows with their piglets inside the pig arcs.

A clay soil such as on this farm where I live is not suitable for outdoor pig production. During the winter, the fields can become very wet; the pigs would have an uncomfortable existence and just make a terrible mess in the field. Sandy soils and chalk soils are ideal, as they are well drained and don't become waterlogged. These fields would support 25 sows per hectare (10 to the acre), but on less dry soils you would probably keep fewer sows. If you are starting this as a new enterprise, I would suggest fewer sows per hectare and see how you get on with them. You can always add more at a later date or take more away. However, if you live in a Nitrate Vulnerable Zone (NVZ), you will need to check with DEFRA what you can and can't do. For example, you must limit the spreading of manure and total nitrogen per hectare in each calendar year averaged over the area of your farm. This amount includes manure deposited by grazing animals. Outdoor pigs would produce a lot of manure, so it is important that you carry out the necessary calculations. Pig faeces and urine deposited on the field could create high levels of nitrate and phosphorus, and you need to be careful that leaking into ditches, brooks, streams and rivers does not occur. Pollution is serious: nitrates can pollute drinking water, and phosphorus can cause growth of algae in streams and rivers, which can lead to a lack of oxygen for aquatic life, including fish, which could result in them dying. A farmer in my area was taken to court and prosecuted for allowing slurry to get into the river that bordered his land.

Soil erosion can also occur, especially on sloping sites. Soil being washed away is bad news for the arable farmer, because it is the top soil that contains the main plant nutrients.

Outdoor pig production is popular with arable farmers because they can grow much of the feed such as wheat and barley; fields can be rotated so the sows and piglets would live in one field for 2 years, and then they would be moved on to another field. A crop such as winter wheat sown on the old pig field would follow, followed by wheat again the next year or another arable crop. You would probably go back and keep pigs again in the original field but certainly not before 2 years have passed. Don't put them onto new grass seeds, as the seeds would soon be churned up. The grass would be better being sown for at least a year before putting the pigs on it, when it would have a solid base. Pigs of course love to root up turf and find things to eat in the soil.

Sows have huts or arks to sleep and rest in. On arable farms, the field would not be fenced, and so an electric fence would be used. The pigs would have plenty of straw in winter months, and a straw door mat would help to keep the huts clean and dry. Baby pigs can die in the cold of hypothermia, and in the summer pigs can get sunburn and heat stroke, and so erecting sunshades plus having a mud wallow in the field is worthwhile.

You obviously need water in the field – this can be supplied out of mobile water tankers or a mains pipe taken into the field. You need to think about lagging of water pipes in the winter, and if you do have mains water supply, pipes must be buried at least 1 metre deep to prevent freezing.

Feed has also got to be taken to the pigs – this can be stored under cover in the field or transported from the farm buildings. Automatic feeders that throw pig nuts from a bulk trailer will take the backache out of feeding.

Pigs that live outdoors probably get fewer respiratory diseases, as they are not living in the more crowded intensive conditions. The first-time farmer may want to just try a couple of sows on this system or go in at the deep end with more but get some experience on an outdoor unit first.

Organic Pigs

Pigs will need to be kept as free range on organic land and fed organic feed. You can't go organic overnight. The farm has to go through a conversion period. The EU deem that you must abide by tighter rules and regulations than conventional free-range pig units. The Soil Association is a governing body that you may want to register with. The Soil Association requires pigs to live in family groups with access to fields if the weather and soil conditions allow. This means they will live outdoors with pig arks, but they are allowed inside housing in bad weather. The Soil Association has banned nose ringing, tail docking and farrowing crates. It also recommends that once you have kept your pigs in a certain field, they don't come back to that land for 4 years. Organic pork will bring you a better price than commercially reared pork; it is best sold at farmers' markets, at the farm gate or in specialist butchers shops. Before you think of starting organic pig production, check out all the EU rules and regulations, and all the up-to-date Soil Association rules.

Pig Breeds

When I was a boy, the three most popular breeds of pig were the Large White, Landrace and Welsh. They are all white-coloured pigs.

The Large White originated from the old Yorkshire breed. They can be identified by their slightly dished faces and erect ears. They grow well in intensive indoor pig units and outdoor systems. When crossed with other breeds, they will produce good-quality pigs that are very uniform, lean and ready for the butcher in a short time.

The Landrace has a very long body with dropped ears. They are excellent for producing pork and bacon, and can be kept on intensive indoor systems or outdoor systems. They produce progeny with a well-fleshed, lean carcass, and they rear large litters. The piglets grow fast and have a good daily weight gain.

Nowadays, commercial pig units usually don't have pedigree pigs, but they have a mixture of breeds, crossbreeds or hybrids that grow fast and produce large litters and generally do better than the pure breeds.

Pure-bred Large White and Landrace (and also the Welsh) have become much scarcer. In fact, the Large White (British Pig Association (BPA) registered) has become a rare breed.

It is certainly easier in this area to find pedigree Gloucester Old Spots and Tamworths than it is to find pedigree Large Whites and even Landrace.

This gilt is a mixture of breeds – Large White, Landrace and Meidam. It shows the upright ears as seen in a Large White. It would be an excellent gilt to purchase.

If you want to run a commercial unit, then crosses such as Large White, Landrace and Meidam are excellent for your gilts and sows, and Large White with some Pietrain make good sires.

If you want to specialise in a breed or a rare breed and sell pork, bacon or both, you must choose your breed. You may become an enthusiast, show them and sit on the breed society committees and eventually become a pig judge. The Rare Breeds Survival Trust publishes a list of all rare breed farm animals.

Here is a watchlist:

Categories	Registered breeding females	Breed
Vulnerable	200–300	British Lop, Large Black, Middle White
At risk	300–500	Berkshire, Large White (BPA registered), Tamworth, Welsh
Minority	500–1000	British Saddleback, Gloucester Old Spot, Oxford, Sandy Black

Other pig breeds include the Duroc, Hampshire and Mangalitza. I have chosen four breeds to describe: Berkshire, Gloucester Old Spot, Tamworth and Welsh.

Berkshire

The Berkshire is the oldest recorded pedigree breed of pig in Britain. It is an attractive-looking pig, black in colour with a white blaze on the face, white socks, white tail and pricked-up ears.

Berkshire sow with 4-week-old piglets.

These pigs are ideal for the hobby farmer or part-time farmer, and they do well indoors and outdoors. They are a favourite of mine because they are usually good-tempered.

They are not as fast-growing as modern hybrids, but you would expect them to be 60–70 kg at 3–4 months old when they will be slaughtered for pork, which has an excellent flavour. They are excellent mothers; the litter size averages 9.98 piglets.

Gloucester Old Spot

This is a popular breed with small holders and pigs kept on a small scale. Gloucester Old Spot pork and sausages are high quality and are popular at farmers' markets, and some butchers specialise in selling meat from this breed.

Gloucester Old Spot sow with 1-week-old piglets.

It is a large breed, white in colour with black (not blue) spots. These spots should be clearly defined. The breed had heavily dropped ears and were once called the 'orchard pig' because many lived in apple orchards, and they would eat the windfall apples. They have large litters, are good mothers and are quiet and easily handled. They are excellent to keep on outdoor systems, since they are hardy and can stand the cold, but they do need a warm hut to sleep in. However, pure-bred Gloucester Old Spots would not be kept in large numbers on commercial intensive pig farms because Large White and Landrace crosses would outperform them, and the Gloucester Old Spot enjoys the outdoor life.

Tamworth

Tamworth boar 12 months old.

The Tamworth is the oldest pure British breed. They are red in colour and have long snouts and upright ears. The sows make excellent mothers. They can produce carcasses for either pork or bacon. Numbers became very low after the Second World War, as farmers kept the faster-growing breeds with larger litter sizes such as the Large White. During the 1970s there were only 17 boars left in the UK. Numbers have now increased, but the Rare Breed Survival Trust has got them on the 'at risk list.' The British Pig Association survey in 2011 showed 493 registered sows. They are hardy pigs and do well outdoors. However, if you keep this breed, you must get organised and sell it as rare-breed Tamworth pork, kept in a friendly way – outdoors with little stress. You need to sell joints of pork and sausages at a farm shop or at a farmers' market. They are a specialist rare breed, but don't forget that they are slower to grow and produce smaller litters than commercial pigs.

Welsh

Welsh sow with 5-week-old piglets.

The Welsh breed was first recorded in the 1870s. They became a very popular breed of pig indeed and one of the major breeds in Britain. This is one of the three major breeds that hybrids have been developed from, the other two breeds being the Large White and the Landrace. The hybrids have done so well that the Welsh breed has declined and is now classed as a rare breed. The pedigree Welsh pig is white with lop ears meeting at the tips, and the body is described as pear-shaped. They are excellent mothers. The piglet mortality is low, and they have a fast live-weight gain. They grow well on indoor and outdoor systems.

During the 1970s and 1980s, I purchased Large White × Welsh weaner pigs on behalf of Higham Lane School. I reared up four per term. The meat was always lean and very tasty.

Chapter 15

Pig Diseases and Ailments

There are so many pig diseases, including some fairly new ones, and so many of them have very similar symptoms. The more pigs you keep, the more experienced you will get in spotting problems and identifying which disease you have present. I think it is hard to diagnose pig diseases; no farmer has seen every pig disease – we are all still learning.

Many diseases can be vaccinated against, so talk to your veterinary surgeon and decide on a vaccination and medication programme that will be suitable for your farm.

Four-week-old weaner pig receiving a vaccination for PMWS.

Don't forget that disease can come onto your farm from new pigs that you buy in. They may look perfectly healthy when you buy them, but they could carry disease onto the farm. Birds such as starlings and waterfowl can also transmit disease and can infect ponds and watercourses. Dogs can spread disease, pig food can be contaminated, and dirty conditions will help disease to spread; dirty shovels and brushes and even tractor tyres used from one pen to another can also spread disease.

Then, there is poor ventilation plus variable temperatures, poor insulation, high levels of ammonia and carbon dioxide, overcrowding, damp and wet bedding, secondary bacterial infections and stress. Keep stress to a minimum, especially when moving pigs and carrying out stockman's tasks. I must stress that good ventilation is very important; keep dust levels low, don't overcrowd them, and keep pens and bedding clean. Housing with continuous throughput of pigs with no rest between batches of pigs will not help; nor will abrupt changes in their diet.

Farmers have said to me 'That vaccine didn't work.' The trouble is, a certain strain of a disease – perhaps a new strain – might not be covered by the vaccine or, more simply, the vaccine could be out of date, might not have been kept at the correct temperature or might have been administered wrongly.

There are over 140 pig diseases and conditions, so it is not possible to look at all of them in this book. I have dealt with some of the main ones briefly, but if you have pigs that are ill, take veterinary advice.

Bacterial Diseases

Enzootic Pneumonia (EP)

EP is the most common respiratory disease in Britain, affecting pigs of all ages. It affects the lower part of each lung. Look for heavy breathing, coughing, high temperature and dehydration, and high mortality.

Antibodies in the mother pass to the piglets via the colostrum, but after about 7 weeks or so the protection disappears, and the young pigs could start to show symptoms. It is a respiratory disease, so pigs don't want to be overcrowded. They need light, airy pens with low levels of dust and ammonia.

Acute disease is usually only seen in new outbreaks of the disease. It is often caused and transmitted by carrier pigs, or it can come in on the wind. The incubation period is 2–8 weeks before symptoms are seen.

Medication can be added in feed and water. A home mill and mix licence is required to put certain additives in the food, but if the feed mill puts them in, you will not need a licence. Injections such as penicillin or oxytetracycline can be given. There are other drugs, so consult your veterinary surgeon.

Erysipelas

Pigs have a fever and a poor appetite, but the most obvious symptom is raised reddish patches in the skin, which are diamond-shaped. Erysipelas is not usually seen in young pigs. Pigs will sometimes go lame and may get meningitis. It is a zoonotic disease, which means that it can be transmitted to humans.

It comes on quickly, and the first sign may be death. The pig will have a high temperature and not want to get up. The disease may cause abortion, stillbirth and mummified piglets. The bacteria are carried in the tonsils, and they can be spread through the faeces or mucus from the mouth. The disease is also carried by birds. It is a soil-borne disease, and the bacteria can live in the soil for many years.

I first came across this disease when I worked full time on a farm after leaving school. The pig recovered after antibiotic treatment (penicillin), and I saw it again when I kept pigs at school when it was also treated successfully.

Vaccinations are available, and prevention is better than cure. Medication such as amoxicillin can also be placed in the water, and feed medications using phenoxymethyl penicillin can also be used.

Glasser's Disease

This is seen mainly in piglets, weaners and growers, but is rare in adults. A respiratory disease in the acute form, it attacks the joints, intestines, heart, lungs and brain. The animals may also have a high temperature, no appetite and a short cough, and may become lame with swollen joints. At the same time, growers can get other complaints, such as pneumonia and pleurisy.

In the chronic form, piglets and young pigs will be slow-growing, and pigs may die. A number of drugs will treat the disease, including penicillin; creep rations can also be medicated, and vaccines are available.

It is a difficult disease to identify when the pigs are alive because so many conditions and diseases are similar, and symptoms are very variable. Characteristic postmortem findings will confirm the disease.

Mastitis

This comes on quickly 6–48 hours after the sow has farrowed. The udder becomes hot and swollen, and the milk is yellow with clots. The mother may not let the piglets drink, and the piglets may starve. It is a good idea to do a pre-farrowing check for heat/lumps 4 or 5 days before farrowing. Mastitis is treated with antibiotics.

Streptococcus suis

This is a widespread disease that can cause meningitis. The disease occurs 2–3 weeks after weaning and may attack pigs up until they are about 4 months old. Many pigs become carriers. *Streptococcus suis* is a zoonotic disease, which means that it is an infectious disease that can be transmitted from non-human animals to humans.

Growing pigs can shiver, and their hair will stand on end. They may froth at the mouth and have fits and jerky movements of the eyes with the head and neck craned back and paddling of the legs. They may develop abscesses. They can be in a lot of pain and won't get up. Sadly, they are often found dead. The disease is spread by aerosol infection or noses touching.

To treat, isolate the pig, keep it warm and give it lots of water. It may not feel like drinking, so connect a hosepipe and slowly get water into the mouth that way, or you can use sports bottles and squirt water into the mouth. Repeat after about 2 hours. Penicillin is a good drug to treat with – but act very quickly. You can inject with a short-acting penicillin. Affected animals don't eat or drink, so injection is really the only way to treat them.

Swine Dysentery

Swine dysentery causes inflammation in the large intestines, and a sign of the disease is mucous and blood in the faeces (dysentery). It is found in piglets, weaners and growing pigs but is uncommon in sows. Other signs of the disease are tail twitching, slow growth and some loss of appetite, the skin turns slightly red, and sometimes the pig will die. Symptoms appear after stress.

The disease is spread by carrier pigs, and pigs can pick the disease up from infected faeces. It can come onto the farm in many ways including humans carrying the bacteria on boots, shoes, tyres, manure forks and shovels. Animals such as birds, dogs, mice and rats can spread the disease. Postmortem and lab tests will confirm the disease.

The disease is such a huge issue for commercial pig farms that pigs that have been affected with swine dysentery are best not kept. The disease is not notifiable, but for the benefit of other producers in your area, you might want to notify them if your pigs get it. Any treatment serves only to keep symptoms at bay, not to cure, so culling is best, and once the animals have gone, strict disinfection is required.

If your pigs contract swine dysentery, your options are as follows:

1. Cull all animals, and when they have gone, extreme disinfection is needed. Then, repopulate cautiously after the buildings have been rested.
2. Lifelong use of an antibiotic called Tylosin (often in the feed) is advised to keep the disease at bay. With Tylosin, pigs should survive to slaughter weight.

Viral Diseases

Congenital Tremor

Newborn piglets are sometimes born with their bodies shaking and trembling. They shake when they are awake and walking, and not when sleeping. The piglets may also sit like a dog. They will probably not be sucking milk and so may starve. They tremble because there is damage to the nervous control system. Piglets can show no signs of the disease after about 4 weeks, but many may die.

The most common cause is infection of the gilt or sow by viral agents in the first half of her gestation period. Rarely, it is caused by an inherited condition involving the Landrace and Saddleback breeds. There is no treatment for the affected piglets, and prevention of the disease is difficult, as the sows show no signs of the disease.

Porcine Circovirus Type 2 (PCV2)

This disease is devastating for the pig industry. It affects weaned and growing pigs. The virus causes two conditions – Post Weaning Multisystemic Wasting Syndrome (PMWS) and Porcine Dermatitis and Nephropathy Syndrome (PDNS). The pigs show signs of pink skin lesions and may waste away slowly starting at about 6–8 weeks; they become emaciated and die. They could have diarrhoea and have paling of the skin, or have slow and difficult breathing.

Pigs can pick the disease up from infected faeces, and this is another disease that can spread by wheelbarrows, shovels, wellington boots and much more.

Drugs to kill bacteria are ineffective (this is a virus). Secondary infections may be controlled using other drugs. No vaccines are available, but hopefully there soon will be.

Porcine Parvovirus (PPV)

This virus causes 'SMEDI' – still birth, mummification, embryonic death and infertility. There are no other signs of ill health, and the disease is seen usually in gilts or young sows and not found in weaners or growers.

Once the pig has had the disease, it will become immune. It is a common disease found all over the world. It can live outside the pigs in buildings for months and it is hard to kill with disinfectants.

There is a combined vaccine against parvovirus and erysipelas available. This is advised, since there is no other treatment.

Swine Influenza

Pigs with 'flu' develop coughing and pneumonia, high temperatures and loss of appetite, and abortions can occur. The heat period may be delayed. The flu can spread very quickly, but thankfully most pigs survive. Carrier pigs as well as wild birds such as ducks can be a source of infection.

There is no treatment for the virus, but secondary bacterial infections can be treated with antibiotics, either injected or given in the water. The pigs will not eat, so the medication is not put in the feed. If pigs are drinking, you can give them aspirin/paracetamol in the water.

Transmissible Gastro-enteritis

In this condition, piglets get diarrhoea, which is watery, and because of dehydration and changes to the electrolyte balance, most will die if under a week old. The disease spreads very rapidly. Sows are also affected with diarrhoea and vomiting.

Porcine epidemic diarrhoea is a similar disease, but there would be fewer deaths in suckling piglets.

The virus is found in the faeces, and if a carrier pig is purchased, the other pigs may catch it from its faeces. Dirty boots, wheelbarrows and machines all spread the disease, or the feed can be contaminated.

The virus likes to live in the cold but is destroyed by sunlight in a few hours.

Most vaccines are not very effective for this disease, and antibiotics will not kill it, but they can be used to reduce secondary infections.

Deficiency Disease

Anaemia

Pigs are born with limited supplies of iron; anaemia is a deficiency of iron in the blood and can affect piglets during the first 3 weeks of life. Sow's milk is deficient in iron, and so indoors the piglet has no way of getting enough iron until it eats creep feed. In the wild, the piglets will get enough iron from the soil. On the first day of life, most piglets are injected with iron to prevent anaemia.

Fungal Disease

Mycotosilosis

Cereal crops that are diseased with fungi produce chemicals called mycotoxins. Pigs will become ill if they eat mouldy and damp cereals that have been bought in; perhaps the feed has turned mouldy in the feed bin because of water leaking in or condensation; the food has been bridging in the bin and has remained there a long time; or there could be a build-up of mould in the other equipment.

Different species of fungi will give different clinical signs, but a number of symptoms may appear, including poor growth, jaundice, respiratory problems, abortion and sows producing no milk. If feed is suspected, send samples to a laboratory for testing, stop feeding it immediately and make sure the pigs are put onto a diet with no mould – that is the best treatment.

Maintenance of feed bins and pig equipment is good practice, and never buy in mouldy or damp grain.

Poisoning

Salt Poisoning

Pigs are in big trouble if they run out of drinking-water, and if this happens an imbalance of salt and water occurs. Salt in the diet becomes toxic, and the salt content of the brain becomes concentrated. It can be a problem for all ages of pigs.

Early signs are the pigs looking unwell, going off their food and being thirsty. As the problem goes on and gets more serious, they will start to be dehydrated, twitch their ears and stand with their heads pressed against a wall. They may sit like a dog and hold their heads backwards, fall over backwards, have convulsions, show signs of meningitis and collapse.

When the pigs are rehydrated, another problem occurs. The brain draws water into it because of the high concentration of salt in the brain. The brain then swells, leading to cerebral oedema, and so you must introduce water gradually in small amounts. Electrolytes in the water can also help rehydrate the pig.

You must ensure that your pigs have water at all times. Watch out for frozen water pipes, and also, if someone has completed repairs to plumbing work, make sure the water is turned back on. Pigs can die within a few hours.

Other Diseases

Scours

Scours is a very common disease in piglets, is also found in weaners and can be life-threatening for sows. Antibodies from the mother's colostrum help, but piglets can still get scours. The condition is caused by a number of agents, not just viruses. The scours will often be white or yellow in colour and have a smell all of their own; the piglets may also vomit and become dehydrated, and eyes may become sunken.

In weaner pigs, the scours can be a range of colours; pigs will dehydrate, lose weight and have sunken eyes, and some may die.

To treat scours, feed can be medicated and piglets dosed orally. Check with your vet for which drugs to use and how much to use. Take faecal samples and send to the laboratory to see what is causing the scours. If it is scours caused by a virus, antibiotics won't be of any use. Make sure pigs drink plenty of water and give electrolyte solution.

I find it difficult to recognise pig diseases. Their diagnosis may need a combination of clinical signs, postmortem examination, and the known presence of the disease in the herd, i.e. if you have seen it in the herd before and recognise it again. Serological sampling for fluorescent antibody tests (FATs) plus other laboratory tests may also have to be carried out. Animals may need to be isolated.

Parasites

Internal

Roundworms including large roundworm (*Ascaris suum*) are often found in the faeces of sows and growing pigs. It is a very common parasite indeed. Affected pigs will cough and will be slow to put on weight. The larvae can cause 'milk spot' characterised by marks on the liver, which means the livers will be deemed contaminated at the abattoir. Female worms can produce a million eggs in a day! There are many medicines available to treat parasites – take veterinary advice for up-to-date treatment options.

External

Sarcoptic mange, sometimes called scabies, is caused by one of two mites – *Sarcoptes scabiei* and *Demodex phylloides*. Infected pigs will shake their heads and rub themselves against the walls of the building, rubbing away the hair. An allergy may occur, causing red pimples to appear on the skin. Lesions can develop. Different drugs are available to treat mange, including Dectomax injection.

A man was driving quite fast along a winding lane. A lady driver travelling in the opposite direction shouted 'Pig!' as she passed. He instinctively saw red and shouted back 'Bitch!.' Around the very next bend, the man crashed his car into a pig, killing the pig and doing much damage to his car. Sadly, he was taken to hospital.

Chapter 16

Other Livestock: Heavy Horses, Goats, Alpacas, Deer and Water Buffalo

Heavy Horses

These horses are a lot of work; they need a large strong stable, some hay, and root crops such as carrots. If they are working hard, they may need horse nuts and oats.

The gestation period is about 11 months; oestrous occurs 7 days after the mare has foaled and will last for 5–7 days at intervals of 20–21 days. When foaling is close, the udder swells, waxy drops appear at the end of the teats, and muscles become slack around the tail head.

Foals are weaned at 4–6 months old, colts are castrated at 1 year old, and breaking in takes place at 2½–5 years old. If broken in at 2½ years, it is for light work only. Heavy horses do not fully mature until they are 7 years old.

Following European Commission legislation, all horses in Great Britain require a passport – contact DEFRA for details.

Heavy horses no longer find a place on most farms, and yet before the invention of the tractor they were essential. The first-generation farmer needs to concentrate on setting up a business, keeping livestock and growing crops, and will have little time for heavy horses. However, if you want to be a heavy-horse enthusiast, this is a wonderful pastime and can only be encouraged.

Breeds

Percheron

The exact origins of the Percheron are not known for certain, but this heavy breed of horse was present in the Huisne river valley in France by the 17th century. They are grey or occasionally black in colour and intelligent, and have a very docile nature. This good temperament makes them easily trained for farm work.

Percheron mare.

Shire

Horses replaced oxen on farms in the 18th century, and these horses were recognisable as Shires. Shires are patient, hardworking, strong horses. The stallions can be black, brown, bay or grey in colour, but they should not have white patches on the body or be roan or chestnut. The mares and geldings can be black, brown, bay, grey or roan. Stallions should stand 17 hands upwards and mares 16 hands upwards.

Shire mare.

Other Breeds

Clydesdales originated in Scotland and are smaller than a Shire, with stallions averaging 16.3 hands. The popular colour is dark brown, with some roans, and they show more white on them than shires. Suffolks are always chestnut in colour and are famous for their clean legs (no long hair on them).

Goats

Dairy Goats

Goats are kept for milk, which can be made into butter, cheese and yoghurt; and meat and fibre.

Many goats are kept by small holders, in just ones or twos, and there are people who live on housing estates who just keep one in the shed and on the lawn (with their neighbour's approval); these goats would be kept for milk or as pets.

Many small holders are very enthusiastic, milking their goats twice a day, rearing kids with the young nannies sold as breeding stock, or keeping them as replacements and selling the billies for meat.

I did exactly that at school, keeping two pedigree British Alpines, selling the milk to staff and parents, and making some cheese. The males were all sold privately for meat. One of my students, Mark, was very keen indeed on looking after the school goats; he persuaded his parents to let him keep a nanny in the garden shed, which he constructed, and she took over most of the garden; this was a normal town house. If you keep goats in the back garden, you must check the deeds on your property because it may say no livestock. Pygmy goats are a very small breed and are often kept as pets.

At school, I built a goat shed with the help of students before and after school and at the weekend. We constructed it in timber, making it in sections and then bolting the sections together. A galvanised metal roof was fitted. The goat shed was light with plenty of air flow but not draughty at goat level. It was divided into two sections, one for each goat, with about 4 square metres of floor space per goat and a pen in front of the goats so that we could look at them. It had hard standing outside and was large enough to store some hay. Goats do like to see each other, so remember this when constructing individual pens.

I do not think it is a good idea to store your concentrate feed next to the goats, because if your goat gets out, it may get to this feed, eat a great amount, bloat and possibly die, so the concentrate feed needs to be stored separately or in a bin of some sort that is impossible for goats to open. Our milking area was in a separate shed that joined on to the living quarters.

I and my students used to let our goats out into one of our paddocks in the daytime. Goats eat grass but prefer the branches and leaves off the trees (we made sure they were not poisonous) but when it came to rain, like most goats, they got upset and wanted to come in. They also came in every evening, spending the night in the shed.

The school goats were fed a coarse ration (for sheep) at the same time as they were milked, which encouraged them to come into the milking shed and stand to be milked. Goats also like green food to eat, such as cabbages and root crops. In their living quarters, they should be fed hay.

It is not worth having a milking machine for just a few goats, as they are easily milked by hand. Goats should be vaccinated for clostridial diseases. It is also not worth keeping a billy goat for just a few females; also, they certainly smell in the breeding season.

Signs that the nanny is on heat (or oestrus) include the following: she wags her tail, she may have wetness around the vulva, and she will bleat a lot. If you get a cloth and rub it around the billy goat and then hang the cloth in the shed (a billy rag), this may encourage her to come on heat. She will come on heat in the autumn every 21 days for just 2–3 days. Most good breeders do not put them to the billy until they are 18 months old.

If you do not have a billy, you need to find someone with a stud, or you can purchase semen. The British Goat Society runs approved AI courses for you to attend, and these are held in the UK. You should clean the pen before kidding and keep it clean with plenty of clean straw. You should be all ready for kidding, with a stomach tube and colostrum on standby in case a weak kid is born.

Goats will bleat or 'talk' to her unborn kids, which is one sign of kidding, and she will become restless, lying down and standing up and pawing the ground. I always take the water bucket out and do not put it back in until she has finished giving birth.

Most births are normal (I have missed many kiddings). The gestation period is 151 days. The British Alpines at school were very, very predictable and did actually give birth on the 151st day and sometimes even at the same time they were served, i.e. served at 4.30 pm and kidded at 4.30 pm, 151 days later, but not always, and not all goats are as predictable as these.

Once the kids are born, you have to dip the umbilical cords in iodine to protect against joint ill, and make sure they drink colostrum. If the kids are to be taken away for maximum milk production, it is usual to leave them with the mother for 4 days and then for them to be taken away. They can of course be left with the mother and left to suckle when they want to. Male goats should be disbudded by the vet, and a rubber ring can be used to castrate them. I could never make any money out of them – by the time we had fed them, they probably made a loss.

Concentrate feed should be increased when milk yield increases. The kids will need milk for at least 3 months. Goat's milk is very good for you, and is said to be exceptionally good for allergies especially eczema. Contact the British Goat Society for information on goats.

I caught Tim (one of my students) feeding the milk back to Anna, our goat, just after milking her. It was a cold January morning, and she could not drink it fast enough; it was lovely and warm. 'What are you doing feeding the milk back to the goat?' I asked, 'We could be selling that at 25 pence a pint.' 'I am sorry, Sir,' he stuttered, 'I got a bit of muck in the milk, so I am just running it through her again for you.'

As an alternative to just keeping a few goats, there are commercial goat farms in this country. The goats are housed in large covered yards bedded with straw and are milked in a purpose-built parlour with machines. The milk is stored in bulk tanks. Milk can be sold to a processor or sold on the farm as milk, cheese, yoghurt and butter. This requires capital to set it all up and a milk contract to sell your milk or clever marketing. In my opinion, you should start with just a few goats, rather than a commercial goat set-up.

Dairy Breeds

Anglo-Nubian

Anglo-Nubian has been developed in Britain by crossing goats based in this country with goats imported from the East. These goats have a Roman nose and drooping ears, and are very tall. They can be many different colours including chestnut brown and black. The milk is high in butter fat and protein, and is good for cheese making. You could expect them to give a milk yield of 4–5 kg in 24 hours.

Anglo-Nubian.

British Alpine

British Alpine.

First recognised in this country at the beginning of the last century, British Alpine has various blood in its breeding which is mostly unknown. A female goat called Sedgenene Faith is sister to the founder of the breed in 1903 – she came from the Paris Zoo. The breed is black with 'Swiss' markings on the head, legs and tail. It is described as 'rangy.' You can expect British Alpine goats to yield about 4 kg of milk in 24 hours.

British Saanen

British Saanen goats on a commercial goat farm.

British Saanen was developed in Great Britain using imported Saanen goats that originated in the Saanen Valley in Switzerland. This is a pure white breed. It is an excellent milking goat with a short coat and a calm nature. You can expect them to yield just shy of 5 kg of milk in 24 hours.

British Toggenburg

British Toggenburg.

This breed has been developed in Britain founded on imported Toggenburg goats. Many are used on commercial dairy farms. You can expect them to yield about 4½ kg of milk in 24 hours. My first goat was a British Toggenburg called Hazel. They are light and dark shades of brown with Swiss markings.

Goats for Fibre or Mohair

The Angora produces fibre or mohair. Heavy fleeces weigh about 6 kg from a male and 4 kg from a female. The fleeces are soft and hard wearing, and can be woven into yarns. Goats are shorn twice a year. Angora goats are a friendly breed suitable for small holders (White Angora wool is produced by rabbits). Mohair is a fibre that can be dyed to bright colours.

Angora buck (male).

Meat Goats

Male goats bred from the dairy herd can be reared up and sold as meat, or there is a meat breed called a Boer, which can be used to serve the low-yielding dairy goats to produce meat. Alternatively, Boer goats can be kept pure and allowed to suckle their kids, which is similar to keeping suckler cattle.

Boer goat.

Goat Diseases

Goats need parasite control and vaccination for clostridial diseases and they are susceptible to many of the diseases that sheep can get, including bloat, frothy bloat and bluetongue, maedi visna and milk fever. See Chapter 13 for further details on diseases.

Alpacas

Alpacas look like small llamas. They are a domesticated species of South American Camelid and are of two types: Huacaya alpaca and Suri alpaca. The Huacaya alpaca are the more popular. They have a short fleece that is closely crimpled, and their heads and faces look like teddy bears.

Huacaya.

Suri.

The Suri alpaca has a long fleece that hangs down in ringlets, like a Wensleydale sheep.

Some people keep breeding herds of alpacas; others keep castrated males, just shearing them for their fibre, and have little worry with them. They are hardy and healthy animals. They love company and are naturally herd animals, so you must keep more than one. A male is called a macho, and a female a hembra; the young are called cria.

They are very expensive animals to purchase, and you should buy from respectable registered breeders only. They like to stay outside with a field shelter, and they do not need high fences like deer. They graze, and you can keep about 12 per hectare (5–6 per acre) on average. They enjoy hay, and you can feed specialist camelid pellets, which are just a vitamin and mineral source.

They should be vaccinated for clostridial diseases, and they need worming every 6 months. Their feet (pads not hooves) need trimming three or four times a year. They will live for 15–20 years and are not slaughtered for meat.

If you do not want to keep an entire male, stud services are available from various breeders. Hembras do not come on heat – when the male mates with the female, the act induces her to ovulate – this is called induced ovulation. Therefore, you can have your cria born when you want them; they are born in the mornings after a gestation period of 11½ months. They usually give birth with very little help.

Some farms keep them in the same field with new-born lambs, as the alpacas will protect the lambs against foxes. They are sheared once a year, before the end of July. The fleece is very soft and is found in 32 recognised colours with shades in between.

Fibre and products are often sold direct to the public – lovely knitwear and wool yarns can command a good price.

Some farms run various courses for you to attend, and The British Alpaca Society promotes education and training for its members and the public. Fleece shows and halter (animal) shows are organised. Details can be found online.

For the first-time farmer, this is certainly different from cattle, sheep, pigs and poultry. However, you need some capital to buy them a paddock, or you can rent land. You will also need a field shelter, and to take courses on keeping them and on how to spin the fibre and possibly make garments. You may need to buy a spinning wheel. They are not for me, but I can see they could be 'on the cards.'

Deer

Red Deer

All supermarkets now sell venison, and it has become popular in many pubs and restaurants. It is low in fat, low in cholesterol and high in protein. Farmed deer need grassland. During the winter months, the deer can be given silage, potatoes and other root crops.

The rut starts in September, when the stags can be dangerous. In spring, the stags drop their antlers. The females calve from mid-May to June. Venison is from animals that are younger than 27 months old; after this age, the animals become tough to eat.

Deer need specialised facilities, catching and holding pens and high fencing to stop them jumping out. To kill them, they are often shot in the field by trained marksmen who hold a firearms certificate to carry out slaughter with rifles; alternatively, they can be taken to an abattoir to be killed.

When handling the stags you can get kicked or gored. Male deer are certainly potentially hazardous. They can get tuberculosis and can carry Lyme disease, which is transmissible to humans from the bite of the infected tick.

Red deer farmed for venison.

The need for specialist handling systems, the high fencing and shooting in the field, and the possibility of getting injured by stags puts me off, so deer farming is not for me and I would not recommend it for the first-generation farmer – at least until you have had experience with other livestock first!

Water Buffalo

'What is the difference between a bison and a buffalo?' 'You can't wash your hands in a buffalo.' The milk from water buffalo is farmed for making mozzarella cheese. Water buffalo found on farms in Britain originally descended from the Asian buffalo. Those on farms now in Britain were first imported from farms in Italy and Romania, and then bred here in Britain.

They are hardy animals, disease-resistant and able to survive on poor pasture. They can also be kept in covered yards, fed hay, straw and cereals, and milked in a parlour. The gestation period is 10½ months.

Water buffalo are kept for meat (which tastes like quality beef) and milk (which is rich and creamy) for ice cream and cheese. To make cheese on a large scale, plenty of specialist equipment is necessary. These are all specialist products and are not as easy to sell at the farm gate as a dozen eggs.

For the first-generation farmer, I would say leave it to the professionals – but after you have kept cattle for a few years, then it is up to you. They are not for me, however, and I would sooner see a good-looking herd of Jerseys or Herefords!

Other livestock enterprises could involve you keeping quail for meat and eggs and also wild boar and ostriches for meat production.

Water buffalo.

Chapter 17

Making a Start with Laying Hens, Pullet Rearing and Broiler Production

I kept my first poultry when I was a teenager living at home with my parents on a private housing estate. I am almost sure that it said on the deeds 'No pigs and poultry'; however, I kept about six cross-breed bantams and some Old English Game bantams. I was told by my parents that I couldn't have a cockerel because of it crowing and upsetting the neighbours. I also came home one day with a Muscovy duck. It took some persuading for me to be allowed to keep her. My parents were worried about the noise the duck would make, but when I explained that the Muscovy was a non-quacking breed, I was allowed to keep her.

I recorded all my bantam eggs laid and sold them. I worked out my profit after the cost of the birds and feed – something I still do today but on a larger scale. My top tip is to keep poultry and layers on free range. In Britain, we eat an average of 186 eggs each per year – I enjoy my eggs, and I must eat over 400 in a year.

Keeping a Hobby Flock

If you just want a few hens to lay eggs for home consumption, you are best to buy hybrids, which will lay plenty of eggs, but if you want eggs all through the winter, you will need extra lighting. If, however, you want to become a fancier or an enthusiastic poultry breeder, you will want to purchase pure-bred large fowl or pure-bred bantams.

There are many breeds to choose from – my wife Sarah has bred and shown Light Sussex large fowl, achieving championships. You need to visit shows, decide what breed you would like, look at and study the breed standards, and contact the breed society and then reputable breeders. Our son, Jonathan, is very keen on the Brown Leghorn – large fowl. He likes them because the cockerels have enormously large red combs and wattles. For his eighth birthday, we bought him a Brown Leghorn cockerel and three Brown Leghorn hens – travelling to the South Coast to buy them from a breeder who breeds some of the best in Great Britain. Jonathan has won numerous championships

Jonathan showing Brown Leghorns.

with them at poultry shows and has bred some excellent birds from them, also winning championships with these.

To breed stock, you will need an incubator or wait for a broody. The eggs take 21 days to hatch. A broody is a hen that has stopped laying, fluffs up her feathers and sits on eggs. She will refuse to get off them when you try to pick her up.

Incubators that turn the eggs automatically are best. If you have an incubator that is manual, you need to turn the eggs three times a day – mark an *X* on one side of the egg and an *O* on the other side. Candle the eggs at 6–7 days and again every 4 days: remove the clear eggs. The chicks will cheep inside the shells before they hatch, which is a joy to hear. You must leave the chicks to dry in the incubator before putting them under a brooder light.

Roseanna transferring chicks out of the incubator.

Free-Range Egg Layers and How to Keep Them

The EU egg-marketing legislation states that for eggs to be labelled free-range, the hens must have continuous access to runs that are mainly covered with vegetation. When I started, I sold my eggs to Bowlers on a 5-year contract and then found my own customers to take them. There are other companies that will take your eggs as well as Bowlers such as Noble Foods. Nowadays, Bowlers offer you an agreement that lasts 56 months. The good thing here is that after signing the contract, Bowlers are legally obliged to purchase and collect all the eggs you produce for 56 months.

Free-range laying hens here at Oak Tree Farm.

One good thing about free-range hens is that they can share the field with cattle, sheep or horses. The main reasons for keeping free-range hens are (1) it provides a good income on a relatively small area of land and (2) it is probably the easiest way of obtaining planning permission for a mobile home, bungalow or house.

Nowadays, you can keep 2500 birds per hectare. The EU welfare directive allows 9 hens per square metre of usable area in the building, and the litter area should account for one-third of the good surface.

Large companies and egg-packing stations these days would not want to bother collecting your eggs if you have only a small flock. You really need about 8000 birds, and these days 12,000 would be the average; however, if you go it alone like me, you can have as many birds or as few birds as you want, provided you have enough hectares and a large enough building.

A new building for 12,000 birds is going to take a great deal of capital. It is not just the building but also all the equipment to go in it. The choice you have got is to spend your money on a large-scale enterprise, which many people have done very successfully (contact Bowlers or a similar company and talk to them and their customers), or you can have a smaller unit and perhaps even use a building you already have on the farm, but before you start, you have got to find someone to take the eggs. You can sell some at the farm gate or on a market stall, but as a newcomer, you need to work out how many you could sell – you would not sell all the eggs from 12,000 birds at the farm gate.

When you have bought your building second hand, employed a firm to build it for you or constructed one yourself, you need to fit it out with all the equipment.

Inside the building, the birds need to be able to walk on slatted plastic floors, which enable the droppings to fall into the pit below. This makes cleaning out easy. The dropping pit is only cleaned out when the birds are sold at the end of lay. In my building, there are two large doors at the far end of the building that enable cleaning out to be done by contractors with a caterpillar tractor.

I have to get organised about 6 months in advance, telephoning the contractors and booking the dates for cleaning out. Before we start cleaning out, we have to dismantle and remove the chain feeder and all the plastic floors. The manure is removed and spread on the farm – it certainly makes the grass grow! Once all the manure is removed, another set of contractors come in and pressure-wash the floors, which are stacked outside, and the inside of the building is also pressure-washed and disinfected. All the equipment is put back in the building before the new birds arrive – this turnaround period lasts 4–5 weeks.

To purchase feed, you will need to telephone a few companies, and feed representatives will come out to see you to give you a price. I have found they rarely give you a price over the telephone. Once you have got a price, you may be offered food for that price for 2, 3 or even 6 months, or you may be offered that price for just 1 month. If you have a choice, you need to make the decision – it's a gamble. I have signed up for 6 months, and the food has gone up and up, and I am happy because I made the correct decision. I have also had the opposite happen, signing up for 6 months, only for the food price to go down An alternative is to go month by month and just agree a price for that month.

Your feed should be offal-, meat- and bone-free, with no antibiotics or growth promoters. When the birds arrive at 16 weeks, I feed a starter ration, which is 18.5% protein with a high level of energy, which includes soya oil. I then change to an 18% protein ration when the birds are about 23–24 weeks and keep them on this until they are sold. The ration includes a yolk pigment, which makes the yolks a good colour. The colour of the yolk is determined by what food the hens eat. Carotenoids are pigments that occur naturally in plants, and the yellow xanthophyll class of these pigments is very important for achieving a dark yolk colour. Birds eating grass and other plants should lay eggs with a good yolk colour; birds indoors with no access to plants will lay an egg with a paler yolk colour. On a large commercial free-range set-up, birds have the opportunity to go outside, but some birds won't go out. To make sure all birds have a really good yolk colour, artificial pigments are placed in the ration. From 40 weeks onwards, I add oyster shell and grit, which helps the shell quality.

On our free-range set-up, feed from the two outside bulk feed bins is taken inside the building via an auger to two bins. A chain sits in a metal trough and passes through the bottom of each bin. This chain is powered by a motor and moves the length of the building in the trough turning the two corners at the end of the shed and then continues back to the bin. It drags the food with it, thus distributing it all around the building. You can set the chain feeder going for as many feeds as you want; a simple time clock in the egg room has pins pulled out – where the pins are out, the chain feeder will work. My birds are fed six times a day using this method, and so no food needs to be carried into the birds manually.

The automatic nipple drinkers are excellent. I used to have bell drinkers, but I found that the birds knocked them, and a great deal of water spilled out, causing wastage and making the manure very wet in the pit below. With the nipple drinkers, you still get some wet droppings underneath, but it is certainly drier than with the bell drinker system.

The automatic nest boxes save a great deal of work, and there is no need for the old-fashioned image of the farmer's wife collecting the eggs in a basket. The birds lay their eggs on a sloping floor, and the eggs roll away on to a conveyor belt. In the egg room, you press the button, and the conveyor belt brings the eggs to you. An operator would collect about 2500 eggs in an hour. We place the eggs on to trays that hold 2½ dozen per tray, but some units have a machine that does this automatically. My unit houses 5000 birds, which means that it is not a full-time job for an operator. On a larger unit, it is possible for one person to collect the eggs from 16,000 birds in a day. However, you may need two part-time workers, as one person may get bored collecting eggs for most of the day.

Fans keep the building cool and automatically go on and off via a time clock. The fans on this farm are set as follows; three fans go on for perhaps 5 minutes and then off for perhaps 5 minutes throughout the day and night (this duration depends on the time of the year and temperature – there is no fixed rule). However, if the temperature rises to 21°C, they stay on all the time until the temperature drops again. If the temperature rises to 23°C, the additional three fans come on and remain on until the temperature drops. On the hottest days and nights in the summer, the fans may run non-stop for 2 weeks or more.

Lighting is also automatic, and the lights also come on and off with the aid of a time clock. When the new birds arrive, they will probably receive only 10 hours of light a day. Then, as the weeks pass, you increase your lighting slowly. I would sooner hold the lighting hours back rather than rush forward – too much light too soon means that you will get eggs earlier, but they will be small eggs, and it seems to take longer to get them to lay larger eggs. If you hold them back, you will get larger eggs from the start. By increasing your lighting slowly, you are imitating spring, and like wild birds such as blackbirds and song thrushes, your hens will start to lay. I provide lighting as follows:

- 16 weeks old – 10 hours' light;
- 17 weeks old – 10 hours' light;
- 18 weeks old – 10 hours' light;
- 19 weeks old – 11 hours' light;
- 20 weeks old – 12 hours' light;
- 21 weeks old – 13 hours' light;
- 22 weeks old – 14 hours' light.

Your lights in the actual nesting boxes and the opening and closing of your nest boxes, which is again all automatic, need to be set up so that the lights go off and the nesting boxes close before the building lights go out. You don't want birds roosting in the boxes because they make the nesting box floors dirty.

Your lighting pattern can cause you problems in the summer months if your building lets natural light in. I have had one problem with the lights. I had 5000 birds delivered and on that day set the lights up for 10 hours – lights on at 8 am and off at 6 pm. This was in mid-July. The lights went out at 6 pm, but instead of the birds going to roost, they moved towards the light, which was getting into the building through the fans. They crowded towards the light and I was worried that they might smother and suffocate. I moved a few hundred back to the centre of the building, but I couldn't move them quickly enough, and they just moved back. I was convinced that many would die and so left the lights on until 11 pm when it was dark outside, and so no natural light seeped into the building. I re-adjusted the lights in the morning so that they came on at 1 pm and off at 11 pm, giving them 10 hours of light. I got up at first light and checked they had not started to move to the light again, but they were fine and settled. This is a real example of hands-on experience – the type of problem you don't find in books.

My birds live on the slatted plastic floors and are allowed on to the scratch area. This has a concrete floor with plenty of wood shavings so they can scratch around and dust-bathe.

Scratch area.

The pop holes into the field are positioned in the wall of the lean-to. These pop holes allow access into the field, and the birds have the choice to go out or stay in. It can be a problem putting your birds back in the building at night – sometimes a few will refuse to go in, but if the lights are on inside the building, and it starts to get dark, the lights will attract them, so set up your lighting pattern to make sure the lights are on when it is getting dark.

Also, if you can give them a feed as it is getting dark, the noise of the chain feeder will attract them in; some farmers even rig up a bell to ring as the chain feeder comes on – like Pavlov's dog experiment. You can now get electric automatic pop holes that will close by themselves in the evening.

The first task the worker or you need to do in a morning is to make sure the fans are running; then check to see if the chain feeder is working, and there is food in the bins, and check all your time clocks to make sure nothing has gone wrong. Check that the water is running right down to the last nipple drinker at the end of the building. As you walk down the building, you pick up any eggs that have been laid on the floor. If you leave a floor egg, another bird will probably lay her egg next to it. When your new birds start to lay, you will get a lot of eggs laid on the floor. Walk round and pick them up at least four times a day – they soon become dirty, which makes them worth less money. In a few weeks, your floor eggs will be down to just a few. Floor eggs will be laid around the perimeter of the building, in the corners of the building and near the nesting boxes.

Sarah working on the automatic egg collection.

Let your birds out into the field – many will not want to go out. Some farmers let them roam all over the field, and some paddock-graze using electric fencing. Paddock grazing means that you can rotate the area, which helps prevent the build-up of disease. In wet weather, the birds may make a muddy mess, and if you can paddock-graze and give the muddy paddock a rest, it will help. Each day you need to check the electric fence isn't shorting out; if it is, it may mean the grass wants cutting.

In the egg room, eggs are placed point down on a tray – 30 eggs to a tray with six trays stacked together; there are 12 stacks on one layer on a pallet, and you place them four high – that is, 48 stacks of six altogether making 720 dozen on one pallet. Seconds are taken out and stacked separately – that is, white-shelled eggs, cracked eggs, dirty and misshapen eggs – however, you won't find every one.

Some egg rooms have a chilled egg store, which helps to keep the eggs at the correct temperature. Eggs should be collected from your farm at least twice a week and obviously sold in strict rotation.

Tasks that need to be completed periodically include cleaning the water drinkers weekly, topping up the litter in the scratch area when needed, looking out for red mite (and if you have got it spraying them) and checking to see if your rat and mouse bait has been eaten.

To make more money you can sell more eggs at the farm gate, and if you buy a grader and grade your eggs, you can sell to shops, restaurants and cafes rather than selling all of your eggs to a packer, so getting more money per dozen this way. My birds will be sold at about 72 weeks of age, which means I have kept them for 56 weeks. During this time, they will have laid at least 300 eggs per bird (aim for 312); 319 eggs is my highest. Most of these end-of-lay birds are taken away to be killed and end up in soups and pies. Some are sold to small holders and people who just want a few to keep in their back gardens – they will keep them for at least another year. On a commercial system, they don't lay enough eggs to be viable, and the shell quality deteriorates, so the birds are sold. Going back to 1970, hybrid hens laid an average of 239 eggs per bird by the time they were 72 weeks old and would eat an average of 127 g of feed per day with liveability at 90.2%. By 2013, genetic improvements have developed the birds to lay 312 eggs; they would eat 115 g of food per day, with liveability at 95.3%. There are many modern hybrids including Hyline, Bovan Brown and Lohmann Browns.

Hopefully you will have made a profit. The biggest problem is the cost of the feed and the price you get for your eggs. If feed prices go up, it doesn't always follow that the price of eggs will go up. If there is a shortage of eggs in Britain, you may get a pay rise, but if there are too many eggs in Britain, you will probably get a pay cut. Your chain-feeder system will probably break down on a Sunday, bank holiday Monday, Easter Sunday or Christmas Day, when you can get no specialist help. This means that you will have to feed the birds manually with buckets and bags. I have also had my mains-water supply pipe cut on three occasions near the village, leaving the hens with no mains water supply. It is a good idea to store water on site. Things also go wrong just as you are going out, and so you have to put them right, and you may end up going out late or not at all!

RSPCA Freedom Foods

To sell to the big supermarkets through a large packer such as Noble Foods you need to be RSPCA Freedom Foods and Lion Code registered. Freedom Food standards are higher than standards set up by the EU egg marketing legislation. They insist that you step up your standards covering the welfare of the birds, feeding and hygiene. The RSPCA will inspect your premises every 12 months. There is a very long list of rules and regulations currently called 'RSPCA welfare standards for laying hens and pullets,' which you can obtain from their website – or you can contact the RSPCA to get all the up-to-date information.

Lion Code

This is run by the British Egg Industry Council (BEIC) and has even more specification than Freedom Foods. As well as the producer identity number, eggs are labelled with the British Lion quality mark and best before date. A BEIC inspector will inspect your flock every 18 months. There are a great deal of rules and regulations, so you need to find out all the up-to-date information by contacting the BEIC.

It is said you get a premium price if you are RSPCA Freedom Food and Lion Code registered. However, you don't need to register with them if you sell your eggs to corner shops, at farmers' markets, restaurants and cafes, and at the farm gate. I am not registered, and I am lucky because I get the same price for my eggs as farmers who are registered, and I know of other egg producers who get the same. I will not register with them unless I have to.

Organic Eggs

These have got to be free-range, and the birds must be fed on organic food. The land must also be managed organically. The EU applies strict rules and regulations, which are more stringent than those for conventional free-range units, and the Soil Association has even more exacting rules and regulations for organic egg production. Rules include the stocking density in the building being lower and the flock size being smaller. Beak trimming is restricted, and more pop holes must be provided to encourage the birds out into the fields. Also, the land needs to be rested.

You will need to check with DEFRA and also the Soil Association to get all the up-to-date rules and regulations. The Soil Association is a governing body that you may want to register with.

Sainsbury's – Woodland Free-Range Eggs

On hot summer days, hens like to spend time in the shade. Sainsbury's have developed a partnership with the Woodland Trust. On these free-range units, the range area for the hens is planted with trees that cover at least 70% of the free-range area. For every dozen eggs sold, 1 pence goes to the Woodland Trust to help funding for tree planting in the UK. You will need to check on all the rules and regulations.

Multi-Tier Systems

If you have more capital to spend, you can invest in a free-range multi-tier system. The birds are still on free range and are allowed out in the daytime, but the inside of the building is fitted out completely differently. The birds can move up or down on different levels, which means that more birds can be kept in the building. Automatic nesting boxes, chain feeders and nipple drinkers are all on different levels with a scratch area on the bottom layer. The manure is removed on a manure belt at least once a week.

A multi-tier system.

Barn Eggs

As the name suggests, the laying birds stay in a building and do not roam outside. The EU directive allows nine birds per square metre. Birds will have chain feeders, drinkers and automatic egg collection, and some will be multi-tier systems.

Enriched Colony Systems

Battery cages have been illegal in Great Britain since 2012, and they have been replaced with a new system called enriched colony cages. These are much bigger than conventional battery cages and are made to house 40–80 hens. They are welfare-friendly and contain perches, nesting boxes and a scratch area. There is enough perch space for all the birds, and the nesting boxes are well used with approximately 95% of the eggs laid in them. The eggs roll away onto a conveyor belt. The scratch area encourages the birds' natural behaviour to scratch around. Nipple drinkers ensure a clean water supply, and a chain feeder brings the food.

Enriched colony system – this cage measures 3.6 metres long × 2 metres wide × 1.8 metres high and houses 48 birds.

Rearing Pullets from One Day Old to 16 Weeks

If you don't fancy keeping laying hens, you can rear pullets instead. You can go it alone – that is, buy some day-old chicks or older birds, rear them up and then advertise them locally, nationally or both. You might sell them all – however, you might not, and when they start laying you will have to collect and sell the eggs. When the birds get over 18 weeks old, they are probably harder to sell, as your customers want point-of-lay pullets, not laying birds. I observed the following on a sign in a farm entrance in Shropshire: 'Chickens for sale, our coop runneth over.'

I would start with just a few birds, say 50, and see if they sell. If they don't all sell, you will have to drop the price, which could mean you making a loss.

A better idea could be that you don't actually own the pullets reared on your farm – you just become a manager for Country Fresh Pullets Ltd c/o Lloyds Animal Feeds, based at Oswestry, or another large company that may be on the lookout for you to rear pullets for them. First, you have to provide the building or buildings – these sheds should be buildings on your farm that are not being used or buildings on a rented farm, or you could purchase a second-hand building. You can't really justify going out and buying a brand-new building if you are a first-time farmer. If you are an established farmer you may be able to justify a new building but even so it would take years to get your money back. Basically, Country Fresh Pullets provide you with the chicks and the feed, i.e. you don't pay for them. You provide the building, the labour to look after the birds, the gas to provide heat for the chicks, the electricity and the bedding (shavings). You don't own the birds – Country Fresh Pullets or a similar company do. They pay you a certain amount of money per bird for rearing them. Chicks are delivered from the hatchery – they will have been beak-trimmed at a day old at the hatchery and vaccinated against Marek's disease; you should be ready with your building all disinfected and set up for the birds.

5-week-old pullets being reared.

The building should have a temperature of 32–34°C, which is lowered gradually until the birds have no heat at all. The lights should be bright, with ad lib feed to start with, and water available. The chicks will be fed on the floor to start with and after a week will receive their food from a chain feeder. A suitable ration for these birds involves starting them on a very-high-protein ration, and then as they grow, the protein level is reduced. An example of a feeding programme that many of my friends use is as follows:

- Premium Starter fed from a day old to 2 weeks – 22.5% protein;
- Super Chick Starter fed from 2 to 4 weeks – 20% protein;
- Chick Ration fed from 4 to 8 weeks – 19% protein;
- Developer Ration fed from 8 to 12 weeks – 18% protein;
- Grower Ration fed from 12 to 16 weeks – 16% protein.

Most of these large companies such as Country Fresh Pullets rear to Freedom Food Standard, and so birds are housed at only 14.5 birds per square metre, but when birds are reared on plastic slatted floors on multi-tiers, you can keep 18 birds per square metre. When the chicks arrive, you give them 23 hours of light on the first day and then drop the number of hours of light – day 2 is 22 hours, and by day 10 you are down to 14 hours, day 13 is down to 11 hours, and after day 13 you keep them on 10 hours until they are sold at 16 weeks.

During their stay with you, they will probably have vaccine placed in the water six times against two strains of salmonella: three times to protect against the Enteritidis strain and three times against the Typhimurium strain. They will also receive vaccine in the water against Gumboro disease at 18 and 25 days old.

I have a few friends who rear pullets as I have just described, but they do tell me that they could do with more money for rearing them! Pullets can also be reared organically – you need to check with DEFRA and the Soil Association for all the rules and regulations.

Rearing Broiler Chickens

Rearing broiler chickens is similar to rearing pullets in some ways. You can go it alone buying day-old chicks rearing them to 56 days old and either killing and preparing them for the table on the farm or selling them live to be killed elsewhere. Broiler chickens are hybrid birds with white feathers. You can keep them indoors for all their lives, or you can go for free range. If you go it alone, I would suggest just starting with a few birds – you need to do your homework and see how many you can sell. It's no good being left with a hundred or so birds that you can't sell – your freezer will only hold so many!

You can always start with a small number, say 25, and then if you sell them easily and get more orders, you can have more next time. You can also create your own website and sell them online.

A better idea, if you have available buildings, is to be a manager. Large companies such as May Park take on pullet rearers. You provide the buildings, the labour to look after the birds, the gas to provide heat for the chicks, the electricity and the bedding. In some cases you own the birds, and in other cases the company owns the birds. They provide the feed, and you will earn money for rearing them.

One of my friends had about 70,000 free-range birds on different farms – producing free-range eggs. He has recently cut down on his numbers of free range and has gone into rearing broilers. The following account details how he successfully rears the birds. He owns a large building that houses 42,000 birds. Chicks are delivered from the hatchery at a day old, and they will have been sprayed with a vaccine for infectious bronchitis. They are both sexes, and they are not beak-trimmed. The building will be all clean and disinfected, and to start the birds off they have a temperature of 33°C. The temperature is decreased more slowly in the winter but by about 3–4 weeks old, the birds will not need heat.

Broiler chicks only 4 days old.

The chicks receive their feed on the floor to start with, and then after 3 days they receive their feed from a pan feeder. They are fed ad lib throughout their lives but are left to eat everything up occasionally. They are fed chick crumbs for their first 10 days and then a grower ration, which takes them up to almost killing age. The grower ration contains a coccidiostat, which is an antiprotozoal agent that kills coccidian parasites. A vaccine against Gumboro disease is put in the water at 18 and 25 days.

The last part of the ration is fed without the coccidiostat – certainly for at least the last 6 days. The birds receive clean drinking-water from automatic nipple drinkers and are housed at a maximum of 38 kg per square metre.

The chicks receive 24 hours of light on the first day; this is then decreased, and after 4 days they will be down to 18 hours of light a day. They will then stay on this number of hours until they are sold.

At 33 days, the birds are 'thinned out': 20% of them are sold, leaving more room for those that are left, and the rest go at 37 days. The birds are taken away to be killed. They weigh 2.2–2.4 kg live weight. Once the birds have gone, it takes a week to clean out, wash out and disinfect ready for the next lot of birds.

How did the frozen chicken cross the road? In a shopping bag of course! If you kill and prepare birds for the oven on the farm, you must contact DEFRA for all the up-to-date rules and regulations.

Rearing Broiler Chickens on Free Range

EU legislation stipulates that the stocking density in the house is lower than traditionally reared broilers that are kept indoors, i.e. 27.5 kg per square metre compared with 38 kg per square metre for standard mainstream producers.

The minimum age for slaughter is 56 days. Birds can be kept in mobile houses, which can be moved onto a fresh part of the field, or they can be kept in fixed-position poultry houses, which will house up to about 20,000 birds. Before going ahead and keeping birds, check up on all the up-to-date DEFRA rules and regulations.

Different, slow-growing breeds are kept on free range rather than the fast-growing birds that are kept on intensive systems. These slow-growing breeds are said to have more flavour. They will need to spend half of their lives outside.

You provide heat for the chicks, starting off with a temperature of 33°C and then decreasing the temperature slowly according to the time of the year but reaching approximately 26°C at 14 days and 22°C at 21 days, and with no heating at all for a few days before the birds go outside. If they are evenly distributed in the pen and active, you know you have got the heating right. If they are panting and sitting away from the heat source, they are too hot, and if they are huddled under the heat, they are too cold.

9-week-old broiler chickens on free range.

Once the brooding stage is over, which is about 3–4 weeks, they must have access to the range – in summer, you can let them out sooner. They will eat chick crumbs to start, then a starter ration, then a grower ration. Finished birds will weigh 2.2–2.4 kg live weight.

You need to watch out for foxes taking your birds and also theft of birds. Rainy weather can cause wet, muddy conditions, and if you have mobile houses you need extra labour to move them. Watch out for disease. If you are rearing birds for a large company, they will probably be sold live. If you slaughter them yourself, you must contact DEFRA for all the up-to-date rules and regulations – don't break the law. Also, you will need to get rid of waste such as feathers and intestines, which means incinerating them or paying someone to get rid of it.

Organic Broiler Production

This is similar to keeping broiler chickens on free range, but the rules are stricter. The birds must live on an organic farm and be given organic feed. They are kept longer before slaughter, and the flock size must be smaller than in free-range production. You may want to rear birds for a large company, or you may rear birds and market them yourself – at the farm gate, at farmers' markets and in specialist butchers shops. There are very long lists of rules and regulations from both the EU and the Soil Association, so you need to check all these before you start.

Corn Fed Chicken

Some birds are corn-fed, which means that a major proportion of their diet is corn, but they need other feed as well to balance the diet. The meat is a characteristic light yellow colour and is sold in supermarkets as well as farmers' markets commanding a premium price.

Breeds of Chickens

If you want to make the maximum amount of profit, keep commercial hybrids such as Hyline or Lohmann Brown for egg laying. For meat production, keep the white-feathered hybrid broiler chickens. You can keep other breeds, but they won't lay as many eggs or be ready for the oven as quickly. However, if you want something nice to look at, or to keep pure-bred birds as a hobby or even become an enthusiast and show them, pure-bred birds are a delightful hobby or pastime. There are many breeds, ranging from large fowl to bantams including Ancona, Araucana, Australorp, Barnvelder, German Langshan, Hamburgh, Leghorn, Marans, New Hampshire Red, Old English Game, Orpington, Plymouth Rock, Poland, Rhode Island Red, Serama, Silkie, Sussex, Wyandotte and many more. If you want to keep a breed, visit shows, talk to exhibitors, read up about them and then decide which breed to go for. I have chosen my favourite breeds below, which are ideal for the hobby farmer.

Sussex

This breed originated in England and years ago was a major breed kept on farms. They lay about 240–260 eggs a year, which are white in colour, and they also make a good table bird. They are found in different colours, including buff, red, speckled and silver, but it is the Light Sussex that is the most common. They are very attractive and white in colour with black stripes on the neck, black tips to the wings and a black tail. You can get Sussex bantams as well as large fowl.

Sarah's Light Sussex large fowl.

Rhode Island Red

This breed originated in the USA. It was a major breed in Britain at one time, but modern hybrids lay more eggs. Many modern hybrids have Rhode Island blood in their breeding. They lay about 280 brown eggs in a year, and they make good table birds. They are a dark red in colour, and you can get Rhode Island Red bantams as well as large fowl.

Rhode Island Red large fowl.

Leghorns

This breed originated in Italy. They make much better layers than table birds, laying about 280 white eggs per year and some birds reaching over 300. They have very large combs: in the male the comb stands upright, and the female comb flops over. There are different colours including black, brown and white. Our son, Jonathan, is very enthusiastic indeed keeping, breeding and showing his Brown Leghorns.

Jonathan's Brown Leghorn large fowl.

Wyandottes

This breed originated in the USA. They are a dual-purpose bird making good table birds and will lay about 200 eggs a year. There are many varieties, including black, blue, blue-laced, buff-laced, gold-laced, partridge, silver-laced and silver-pencilled.

The silver-laced Wyandotte is my favourite, and I kept the bantams on the school farm. The bantams are miniatures of the large, heavy breed. They have white feathers with black edges and yellow legs. They are docile, easy to show and very attractive indeed.

Silver-laced Wyandotte bantam.

If you don't want to show your large fowl or bantams, you can always show your eggs. There are classes at most poultry shows for large fowl and bantam eggs. Classes may include six large any colour, three large brown, one large single, one large contents, three white bantam, one bantam single, one bantam contents and more, which sometimes includes the best decorated eggs.

Roseanna's champion eggs.

Chapter 18

Chicken Diseases and Ailments

There are many diseases of chickens, and many of these have similar symptoms, especially respiratory diseases, where birds have a nasal discharge, cough and sneeze. The more chickens you keep, the more experienced you get at detecting that you have got a problem, but for identifying the disease you may need an experienced poultry person, a vet or a diagnosis confirmation following specialist tests in a laboratory. Postmortems can tell us a lot, and for some diseases blood tests can give a result.

To prevent chicken diseases, good biosecurity is essential. Keep the birds clean, disinfect after each batch of birds on an all in–all out system, or if you keep chickens in the same pen for over a year, clean out and disinfect on a regular basis. Keep water drinkers and food troughs clean, don't let bedding become wet, dip your wellingtons in disinfectant, trap or poison vermin, spray for red mite, keep the building well ventilated, feed good-quality feeds that are not stale, keep stress to a minimum and spend time looking at your birds. Vaccinate against certain diseases. Talk to your vet and decide on which vaccination programme will suit your flock. When you buy your birds, find out which diseases (if any) they have been vaccinated for, as most vaccines work best if given at the hatchery.

The following descriptions of chicken diseases are by no means a full account, but they will give you some idea of what could occur. When you read through the number of chicken diseases, there really are so many with almost the same symptoms, and you could frighten yourself thinking your chickens have got every disease under the sun. This is a difficult subject; keep calm and get expert advice.

Respiratory Diseases

Avian Influenza – Bird Flu

This is a notifiable disease that is highly contagious. It affects not only the respiratory system but also the digestive and/or the nervous system. It is caused by different strains of viruses closely related to the human influenza virus. Signs of the disease are finding dead birds that have died suddenly, birds with diarrhoea, nasal discharge and coughing. The birds will eat less, and egg production drops. A laboratory test is needed to confirm the disease. It is spread by contact with infected faeces, oral contact, contaminated boots, water, feed, clothing and other equipment. A laboratory test is needed to confirm the disease. There are 16 strains of bird flu – the H5N1 strain is deadly. In Great Britain, this strain was first found in a parrot in quarantine, and in April 2006, it was found in a dead swan. Since then, it has been confirmed that birds died of avian flu on a turkey farm in Suffolk. The H5N1 strain has been found in turkeys and also in wild swans. The avian H7N7 strain has been found in chickens. The avian H6N1 strain has also been found on two farms in England, and in November 2014 the H5N8 strain was also found. The virus can infect people, but it is not common. The World Health Organization reported that there had been 583 cases of H5N1 (all abroad) in humans worldwide. If an outbreak of this disease occurs, it has to be dealt with very quickly and efficiently to stop it spreading further. Contingency plans are held by DEFRA. Culling and surveillance are the most effective ways of protecting from and controlling an outbreak.

Infectious Bronchitis

This is a highly contagious viral disease that causes damage to the kidneys. It is probably one of the commonest diseases in commercial egg-laying flocks. The birds sneeze and cough, and egg production certainly drops. Many eggs have rough or soft shells, owing to an infection in the oviduct. The disease is easily spread through the air, by contact from one bird to another or from items such as wellington boots.

Poorly ventilated poultry houses with a high stocking density result in birds breathing in stale air, so it is very important to keep the air flowing. It is the ammonia that changes the mucous membrane of the trachea. If you spot the disease early, send off some birds to a laboratory for a postmortem to confirm you have the disease, and with the correct medication in the water the rest of the flock should recover. Losses may still be high, even with treatment, and production losses can be very severe. If you have to use antibiotics, check to see when you can sell your eggs, i.e. what the withdrawal period is. Vaccination is available.

Mycoplasma (Roup or Common Cold)

This is a very contagious disease. Mycoplasma are somewhere between bacteria and viruses. It is not officially a bacteria but can be killed by some antibiotics. There are different species of mycoplasma, *Mycoplasma gallisepticum* being the commonest. Birds will have swollen eyes with foam in the corners, breathe noisily, sneeze and have a nasal discharge that smells. Adult birds may show no symptoms. Chickens infect one another by breathing the bacteria in, and they can also pick it up from their bedding material or dust. It can also be spread by contaminated footwear, water drinkers and feeders, and wild birds such as starlings can also bring in the disease. If it is found in your flock, the birds are best treated with mycoplasma-specific antibiotics. Contact your vet to supply them and give you dosage details.

Newcastle Disease or Fowl Pest

This is a highly contagious viral disease. The symptoms do vary, but birds may have a nasal discharge, wheeze, find it difficult to breathe, and have cloudy eyes and green droppings. The nervous system is sometimes affected, plus you will get soft-shelled eggs. Young birds usually die, but older birds will recover. There are not many cases of this notifiable disease in Great Britain, but it is found in many other countries. It can be transmitted through the air or can be carried on wellington boots and spread by using dirty equipment. There is no treatment; culling and surveillance are the best ways to prevent the disease. If you had the disease, DEFRA would introduce disease-control measures on your premises and the surrounding area. You would only vaccinate against it if an outbreak of the disease started somewhere.

Other Diseases and Ailments

Acute Heart Attack

In this condition, birds just drop dead. This is sometimes found in fast-growing broilers and pure-bred large fowl, especially large cockerels. You may see birds kick, roll over and gasp, but they die very quickly indeed.

Cannibalism – Feather Pecking

Birds can peck at each other, especially at their vents. They may suffer from anaemia, and in severe cases they can be pecked at so badly that the vent will bleed, which will encourage more pecking, and then the rectum and intestines are pecked out, resulting in death. The bird with all of its feathers is probably the one doing the pecking! On a small backyard poultry set-up, you may need to cull the culprit.

A shaft of light coming into the building will often be pecked at, and I think this could cause the birds to start pecking each other, so block off any holes that let beams of light through. You can also lower the light level, which may help, but once they start feather pecking, it is difficult to stop. I always feed layers' meal and not layers' pellets, because it takes longer for them to eat the meal and so gives them more to do and less time to think about feather pecking. I have experienced feather pecking with the commercial flocks, but beak trimming does make it less common. This, however, is a limited practice in organic flocks.

Coccidiosis

This disease is caused by a protozoal parasite. There are different strains causing coccidiosis, and they live in different parts of the intestine. The different strains will show slightly different symptoms, or different strains may work together to cause the disease. Often the birds are not happy, ruffle up their feathers, stand hunched, close their eyes and have diarrhoea that is sometimes bloody. It can occur when birds are living on wet litter in poorly ventilated houses, and the stocking density is too high. I have seen this disease in my small flock of purebred poultry, and indeed some birds have died. However, many get better with medication in the water such as Baycox, prescribed by your vet. Chicks and young birds are most prone to infection. Day-old chicks can be vaccinated with Paracox. Anticoccial preventarive drugs can be put in the feed. Coccidiosis is very species-specific, and so chicken coccidia never affect sheep and vice versa.

Contact Dermatitis

This is also known as hock burn or pododermatitis. It affects the skin, especially in broilers. Constant contact with litter can cause hock burn, and it is found on the rear of the foot, the foot pad and sometimes the breast. Black marks with lesions, ulcers and abscesses appear, and then the birds may limp. Birds living in crowded conditions with damp or wet litter and poor ventilation are vulnerable. Affected birds should be moved to a clean, dry litter space, but if this occurs on a commercial unit during the last weeks of production, the birds will just be killed.

Egg-Bound Hens

This is usually caused by an imbalance of calcium. On a commercial unit with thousands of birds, you rarely see this because birds will lay their eggs in nest boxes where you can't see them laying. It is often caused by a large or misshapen egg. I have encountered it with my small flocks of pure-bred fowl. The hen becomes distressed, is in pain and strains as the egg is stuck in the oviduct. You should be able to feel the egg or see part of it. If you place her vent in warm water and then massage in olive oil or lubricant, you should be able to free it. Without help, it can be life-threatening.

Egg Peritonitis

Hens with damaged oviducts may suffer egg peritonitis. This is when the yolk is laid internally into the abdomen. A number of yolks will collect, and as the bird produces more and more yolk, she will feel discomfort and will adopt an upright stance. The abdomen may become infected, and birds will often die. Each year, I would see peritonitis in the commercial flock – perhaps just three or four cases. There is no treatment.

Heat Stress

If the building becomes too hot with high humidity and little movement of air, the birds will pant a lot and stretch out their wings. They will drink more water, eat less and lay less. In some cases, when it becomes very hot birds can die; buildings that are too hot in summer need more fans and possibly fewer birds. Fans will need to be on 24 hours a day in hot weather on commercial units – these fans should be thermostatically controlled. When I have been away from the premises, and it comes out very hot, I often worry and hope the fans are working. I usually telephone to ask the workers to put my mind at rest – they have always been fine.

Infectious Bursal Disease (Gumboro)

This is a viral disease that affects the immune system. Lymphocytes are destroyed, and so the birds' resistance to infection becomes poor. It affects young birds up to 4 months old, but it is critical between 2 and 4 weeks. Some die suddenly, and some recover.

Birds will be unsteady on their feet and have ruffled feathers and watery diarrhoea. They will quickly become dehydrated. It is spread orally but may also be spread via the respiratory tract or the eyes and can be spread by contact with contaminated equipment. Antibiotic cover can be beneficial, and vaccination is available.

Marek's Disease

This disease is caused by a herpes virus. Chickens from 6 weeks old can be affected, but it is more likely to be seen in birds that are 12–24 weeks old. The disease causes tumours and affects the nerves. The birds get paralysis of the legs and wings, and lose weight. The virus can survive for long periods out of the birds. The disease can be transmitted through infected faeces and by contact with other chickens, including via secretions from the eyes, mouth and nose. Infected birds need to be culled. A vaccine is available.

Salmonellosis Salmonella: Enteritidis and Salmonella Typhimurium Infections

The infections are caused by bacteria and are associated with human disease – food poisoning. The digestive tract is infected, and eggs can become infected in the fallopian tube. Adult hens can lay infected eggs without looking ill and in fact show no clinical signs. Young birds may droop their wings, have diarrhoea, eat less, close their eyes, ruffle up their feathers and look generally unwell, and there may be many deaths. Boot swabs taken on routine salmonellosis tests (current regulations) are sent to the laboratory to see if you have salmonellosis. Birds become infected through bacteria in their droppings, from a contaminated environment such as from dust and litter, plus dirty faecal contamination on eggs for hatching. Eggs to be hatched can be washed with Virkon in the water to kill bacteria. Point-of-lay pullets purchased from large companies such as Country Fresh Pullets Ltd will have been vaccinated against these infections by vaccines placed in their drinking-water. Pullets purchased from a backyard poultry keeper may not have been vaccinated against these infections. Check with the breeder – you don't want the disease in the poultry, and you don't want it in the food chain. Rats, mice and wild birds can carry the disease. Poison or trap the rats and mice, and don't feed poultry outside, or just give them enough feed outside that they can eat in a short period to discourage wild birds. Salmonellosis can be treated with various drugs, including neomycin and amoxicillin. Consult your vet for the best treatment for your birds. I insure against salmonellosis.

Parasites

External

Red Mite and Northern Fowl Mite

Red mites are very common indeed. They can live in your building for a year without any birds present, and the eggs or oocysts can remain dormant for a year. Red mites prey on the birds mainly at night by sucking their blood. Egg production will drop. Red mite will live in the wooden building during the daytime. Always be on the lookout for them, check crevices in the walls, floors and perches, and pick birds up to look for mites on their skin under the feathers. The mites will crawl on you, which makes you itch, but they won't suck your blood. I remember sitting in a staff meeting at school surrounded by townie teachers, and my arm started to itch. I looked and found red mites crawling up my arm. I disposed of them directly with my finger and thumb and thought the other members of staff wouldn't want to be sitting next to me if they knew!

Red mites are very unpleasant for the birds, so at the first sign of them spray the building with an approved red mite killer such as 'Ficam W,' which is a residual spray. With this product, you spray the building, but you don't have to remove the birds. There is another mite called the northern fowl mite, which spends the whole of the time on the bird. Infected birds may have pale wattles and comb plus dirty patches in their feathers, and in severe cases may develop anaemia and die. For northern fowl mite, you treat the birds with a pyrethrum-based powder.

Scaly Leg Mite

These small mites burrow under the scales of the legs. Raised encrusted scales will form. These mites irritate the birds. To cure, dip the legs in surgical spirit once a week for at least 3 weeks. Don't pull the crusts off. I have seen scaly leg many times.

Internal

Worms

Birds with worms will be slow to grow and possibly waste, and may be found gaping. The worms live in the trachea and intestines. They can often be seen in the faeces, which can be green. As many as six different species can be found, including large roundworm (*Ascaridia*). Rotation of your free-range pasture will help to prevent worms. Birds become infected orally. Flubenvet is the only licensed poultry wormer and is an excellent product.

A lady was sitting on her own in a pub with a bottle of champagne. A man came into the room, ordered a bottle of champagne and sat at the next table. 'Are you celebrating?' enquired the lady. 'Yes,' the man replied. 'What a coincidence – so am I,' she said. 'What are you celebrating?' she then asked. The man looked very happy. 'I am celebrating because I am a poultry farmer, and after many years of trying I am at last getting fertile eggs from my hens. What are you celebrating?' he asked. The lady looked very happy indeed. 'I am celebrating because after many years of trying, I am finally pregnant. What is your secret?' she asked. 'I have used a different cockerel,' the farmer replied. 'What a coincidence – so have I!' the lady replied with a glint in her eye.

Chapter 19

Making a Start with Ducks, Geese and Turkeys

Ducks

Frost in November to carry a duck
The rest of the winter will be sludge and muck.

Ducks are kept for egg laying, meat or both and also for pets and a hobby.

If you want to keep ducks for meat, you need to find out how many you can sell before going out and buying a large number. Duck is not as popular as chicken; your main ways of selling them are at butchers shops, farmers' markets or the farm gate.

If you are going to breed some, you will need one drake for five or six females. It takes 28 days for duck eggs to hatch and 35 days for Muscovy duck eggs to hatch. I have hatched duck eggs under the parent bird successfully, but often they don't make good sitters or mothers. A broody hen will probably make a better job of hatching them, but duck eggs are usually larger than hen eggs, so she will not be able to sit on as many, perhaps about eight.

Broody hens don't wet the nest like a duck. A duck will go and have a swim and come back with water on her feathers. She will then sit on the nest, making it nice and humid. I help in this situation by sprinkling water on the duck eggs on a regular basis, especially for the last few days. If you are going to use an incubator, it is best to get one that automatically turns the eggs; if you are turning manually, then do so three times a day. Mark an *X* on one side and an *O* on the other. Candle the eggs at 6–7 days and then again every 4 days – remove the clear eggs. Leave the ducklings in the incubator to dry and then put them under a brooder.

Our cat, Henry, is a real character and enjoys killing short-tailed field voles, mice, shrews, rabbits and even moles. One day, he was a real horror and killed and ate one of our ducklings. We now call him the duck-filled fatty puss!

Keeping Ducks for Meat Production

If you are keeping ducks for meat, you will probably buy day-old ducklings from a reputable company such as Cherry Valley at Caistor, Market Rasen, Lincolnshire. Ducklings will need heat for about 2 weeks (37.7°C under the brooder and 26.6°C at the side of the brooder) depending on the time of the year, reducing the heat gradually. Start your day-olds on a duck starter ration, which would need to be about 20% protein, and change gradually to grower and finisher rations. The feed will be in the form of pellets. There is no need to provide a pond, but they need water containers deep enough to submerge their head and beaks. The pen needs to be well ventilated and kept clean, and this can be achieved by daily topping up straw bedding. Straw quality is important, as mouldy straw will cause mortality problems.

Broiler Ducks

Broiler ducks are used for production of duck meat, and they can be purchased from Cherry Valley Ducks. The company has developed the Cherry Valley Duck, a hybrid that is a Pekin-type duck that has been bred and selected for fast growth and live-weight carcass quality. They will be ready for slaughter at 5½–7 weeks old, and an ideal live weight is 3–3.8 kg. Ducks are housed at less than 25 kg per square metre, and these days large duck houses would house 9000–10,000 birds. They are not usually vaccinated with anything and do not receive routine medication.

You can become a contract grower and rear ducks for companies such as Cherry Valley. You provide the buildings, labour, heat and bedding, they provide the ducks and the feed, and you get paid for rearing them.

Broiler ducks can be kept on free range and organically. You need to protect them from foxes and other predators such as gulls, which could involve you using electric

Cherry Valley standard commercial ducklings aged 7 days.

Cherry Valley ducklings aged 28 days.

netting. The netting is ideal to make paddocks. Rotate your paddocks, as ducks can make a great deal of mess, especially in winter. To go organic, you need to check out all the EU rules and contact the Soil Association. If you are killing ducks on the farm, you must contact DEFRA to get all the up-to-date rules and regulations.

Keeping Ducks for Egg Production

Keeping ducks for eggs is easy, but you need to find out, or at least get some idea of, how many you can sell before buying lots of laying ducks. Duck eggs are not as popular as hen eggs; they have a stronger flavour, and so many people don't like them. Many ducks are obviously kept to lay eggs, and many ducklings are reared and sold as the broiler ducks. Cherry Valley have an excellent enterprise, producing their own duck eggs.

You can buy your ducks from a firm such as Noble Foods (head office is in Hertfordshire) purchasing them at a day old or older or at point of lay – 18 weeks. The older they are, the more money they will cost. Ducks purchased from Noble Foods would be vaccinated against salmonella between 1 and 5 days old. They are white-feathered hybrid ducks (Hybrid 109) that contain a great deal of Campbell breeding. They will lay about 300 eggs over a year. These days, specialist duck-laying units consist of either mobile sheds that would house about 1500 birds or fixed buildings that may house 2000–4000 birds on free range. Many buy the birds at point of lay, which saves the bother of heat and rearing the ducklings. At 18 weeks, the ducks would need 17 hours of light a day, and this lighting pattern is sufficient until the birds are sold at end of lay. They would start to lay at 24 weeks old and would lay for 60-plus weeks. For younger ducks, feed them on a starter pellet until they are 8 weeks old, then transfer them to chicken layers' pellet. You can also give them some wheat or mixed corn. Don't feed rations containing coccidiostats. Our daughter, Roseanna, keeps a few white Campbell-type ducks for egg production. She uses the farm's poultry meal, which costs her nothing, and she sells the eggs, pocketing all the money to buy toys. She has shown these ducks successfully, winning prizes, but the last time she showed her drake, Sparkle, he bit the judge, bruising his hand and arm, so Sparkle is now retired from showing. Instead, she now shows Call ducks, which are a very small breed indeed, and she has won many championships with no problems.

Egg-laying ducks can be kept indoors in wooden fox-proof houses that are well ventilated all the time, or they can be let outside. Don't let them out until mid-morning when most of the eggs will be laid. Ducks are not the best birds in returning home, so a wire pen around the duck house or electric netting will save you some time. You will need to rotate your paddock areas.

Cherry Valley standard commercial parent ducks producing eggs for hatching – note the white nesting boxes.

Ducks can be kept organically – you need to check EU requirements for stocking density and other rules; also, you may want to abide by Freedom Food rules, Soil Association rules and also a new body called the Duck Assurance Scheme. If you have 50 or more birds, don't forget to register your flock with DEFRA; if you have fewer than 50, it is not a legal requirement, but you are encouraged to register voluntarily.

Duck Breeds

Aylesbury

This is the most famous breed of duck, a heavy breed that originated in Aylesbury, Buckinghamshire. They are not as common as you think; many white ducks look like Aylesburys but are actually hybrids. It is a good meat duck and can also produce up to about 120 eggs a year.

Aylesbury.

Campbell

Originating in Gloucestershire, this is a light breed. They are really very good egg layers, laying 200–300 eggs in a year with some strains achieving 340. They will also sell for meat but would take much longer to grow than Cherry Valley ducks. There are three colours: khaki, white and dark.

Khaki Campbell.

Indian Runners

Indian Runners originated in the East Indies and not India. They stand erect like penguins, and they run instead of waddling like other ducks. There are better breeds for meat however; they can lay up to 200 eggs in a year. They come in many colours including white, fawn, silver, chocolate, apricot, black, blue, and fawn-and-white.

Fawn-and-white Indian Runners.

Pekins

Pekins.

Pure-bred Pekins are a heavy breed that originated in China. They are fast-growing and produce a good, white meat. The breed has been used in developing quick-growing hybrid broiler ducks. Given the choice of the two, I would keep the Cherry Valley hybrids because they are so fast-growing. Pekins are cream in colour. They can lay up to 140 eggs a year.

There are many other breeds of ducks, but these pure breeds are for the fancier or enthusiast to breed, look after and show; they will not make you a living. Other breeds would include Muscovy, Silver Appleyard and Rouen. Call ducks are the toy ducks of domestic ducks and make wonderful pets.

Geese

Keeping geese is very easy indeed. Geese are poor egg layers compared with hens and ducks, and so are mainly kept for meat – especially for Christmas.

Like all the other poultry enterprises, you need to find out how and where you can sell your birds. You could probably sell a good many at market, but this is a hit-and-miss business. You can contact butchers, but many of these will have a good relationship with suppliers that have been supplying them for years, although you can certainly try. You can advertise or sell at the farm gate or at farmers' markets, or create a website and advertise there, and you will meet customers that you have not met before. I think if they order a bird from you, and you don't know them, they should leave a deposit. If you have reared your geese, killed and dressed them, and the customers change their mind and don't collect them, you have to either put them in your freezer if you have room or take them to market as long as the last market before Christmas hasn't gone. It is more difficult to sell a goose in January.

I know a farming family, Richard, Joy and Michael, who have built up a steady relationship with customers and have increased their bird numbers to 75. They purchase hybrid goslings called the Legarth. The Legarth hybrid is bred and sold by a reputable firm called Norfolk Geese at Pulham Market, Diss, Norfolk. You can see the flock from the A5, and we as a family look out for them during the latter part of the year. Our children look sad in January when no geese can be seen in the fields.

Goslings can be bought at a day old, a week old and 16–18 days old, or sometimes older. The older they are, the more money they will cost you. I bought my first goslings in the 1970s and have bred them with natural hatching, i.e. the mother sitting on the nest, and I have hatched eggs in the incubator. Before your goslings arrive, you will need to have your pen clean and disinfected, with food and water ready. The goslings require heat: 36–37°C for the first week, reducing it down to 23–25°C at the end of the second week. If it is warm, you may only need the heat at night during the second week. After the third week, no more heat is required unless it is very cold, which is unlikely, as most goslings are purchased in June ready for killing at Christmas.

A circular pen is better than one with corners because it stops the goslings crowding. You can let them out on to the grass after they are 2 weeks old as long as it is not cold and wet. If it starts to rain, bring them in. After about 5 weeks, they will cope with rain. You can keep about 150 birds to the hectare (68 to the acre). Geese do like water to swim in. Back in the 1960s, my Aunty had some geese – she let them out, and they found the river and swam downstream for miles, so it took a long time to get them back.

Start your goslings off with chick starter crumbs, and then give them a grower ration at about 6 weeks old. They will eat flint grit and more and more grass as they get bigger. After 10 weeks, reduce the feed to just one small feed a day. At this stage, they can have some rolled barley or wheat in the ration as well as the grower meal or pellets. Keep the birds on a reduced diet until the last month or month and a half before killing, then increase the feed up to ad lib – continuing with rolled barley, wheat and pellets. Geese love grass to eat, but they like it short, young and fresh the best. If the grass grows above 12 cm, it is best mown for them.

The Legarth is a very popular gosling to buy, producing an excellent finished oven-ready bird weighing 6–6.5 kg (13–14 lb) at Christmas, with some reaching weights above this. Weights will vary, however, according to how much feed they receive and overall management. This is certainly a hybrid I would keep rather than the pure breeds.

A trio of geese or a few more kept around the farmyard make good guard 'dogs' and will make a terrific noise if you have visitors or intruders. Ganders can get aggressive, especially in the breeding season. When I taught Rural Studies in a secondary school, I kept and bred geese, Gregory being our star gander. He would run at you with his head down and would grab hold of your leg. To stop him, I would walk forward and hold him by his neck and then pick him up and put him under my arm. I would then walk around the classroom with him, looking like Rod Hull and Emu, and because I held him this way he didn't have a go at the pupils. Some pupils were successful in picking Gregory up outside, but the big mistake was to dither and start to walk backwards – he would then grab them.

A few geese kept full time on the farm will produce some eggs, perhaps up to 30 or 40 in each season. They should start to lay in the middle of February and may continue to lay into July or August. These eggs can be hatched in an incubator, or the goose can be left to hatch some. Recently goose eggs have become more popular to eat. They are enormous – you certainly don't want two on your plate at breakfast. Goose eggs are also popular with enthusiasts, who decorate them with wonderful designs and patterns.

If you intend to slaughter birds on the farm, you must check the up-to-date rules and regulations that DEFRA have set. If you don't fancy killing them on the farm, you may be able to find someone to do it for you.

Geese can be kept organically – check the current EU regulations and the Soil Association rules.

Goose Breeds

Hybrid – Legarth

In my opinion, this is a good buy if you want a strain of goose that will 'do well' and hopefully make you a profit at Christmas.

Hybrid Legarth geese.

Pure Breeds

Pure breeds are for hobby farmers; they won't make you much money but may give you a rewarding pastime, which may include exhibiting at shows.

Chinese

These can be kept for meat and eggs. A noisy breed originating in Europe and a descendant of the wild Swan Goose, they are grey or white in colour. They are the best geese to produce eggs, 40–80 per bird in a year. A small goose weighs 3.6–5.4 kg (8–12 lb) live weight.

Chinese.

Embden

This is a large white breed that originated in Germany; it is good for meat but will lay only about 20–30 eggs a year. This breed has been used in developing hybrids for meat production. They can weigh up to 15.4 kg live weight (34 lb). A strain of Embden goose called the Stowe goose is sold by a reputable firm called Gulliver Geese at Stowe Bedon, Attleborough, Norfolk. These geese are 6–6.5 kg (13–14 lb) when ready for the table. They are hardy and fast-growing,

Embden.

and should do well for you. Many white geese are reported and claimed to be Embdens, but most are hybrids. Embdens look good, but to make the most money, I would keep either the Legarth or the Stowe.

Steinbacher

These geese originate from Eastern Germany. Blue and grey are the standardised colours. They have orange legs and feet, and an orange bill with a black bean-shaped mark. Ganders weigh 6–7 kg (13–22 lb), and geese weigh 5–6 kg (11.02–13.22 lb).

Steinbacher.

Toulouse

The Toulouse is a large breed originating from Toulouse in France. They are blue grey in colour with brown bars and some white. There is now also a white and a buff strain. They will lay about 30 eggs a year. They weigh up to about 11 kg (24.2 lb) live weight.

Toulouse.

There are several other breeds of geese including Brecon Buff, African, Pomeranian and Sebastopol.

Turkeys

The turkey is a funny bird
Its head goes bobble-bobble
All he knows is just one word
And that is gobble-gobble.

Approximately 23 million turkeys are reared in Great Britain each year. Breeding turkeys is specialised, and so I suggest rearing them for meat if you want to make money, but you can breed a few for a hobby, and there are a number of attractive old-fashioned breeds to choose from. Keeping turkeys can be a profitable business.

Turkeys are very popular at Christmas – more popular than geese – but you need to find out how many you can sell before you start. Farm-gate sales and farmers' markets may account for some that you can sell for a good price. You can also advertise them in newspapers and on the internet. If you sell to butchers or restaurants, you will probably get less money, and if you take them to market, you could do well or the opposite. Butchers often have regular suppliers that they have dealt with for years, so it is sometimes hard to strike up a business deal. My advice is to start with a small number, and if you sell them all, you can have more the next year. If you advertise, and you have not met your customers, arrange for them to visit the farm and leave a deposit, as they could order a turkey and then not arrive to collect it.

Surplus turkeys can be put in the freezer, and you can hopefully sell them at a later date. The Anglia Turkey Association has members all over Great Britain; they operate a 'turkey exchange,' matching producers to buyers. Alternatively, Kelly Bronze Turkeys at Danbury in Essex provide a franchise business opportunity. Kelly Turkeys may offer you a franchise depending on where you live. If you have near neighbours already rearing Kelly turkeys, you will not be allowed to take part in the scheme, as you could affect their sales; they also take into consideration the ease of access to your farm for deliveries.

To start you off as a new producer, they suggest you have 150 turkeys that are off heat, they will deliver them, and at this stage you don't pay for them. They also provide the feed; you provide the buildings and labour.

You grow them to the finished weight, and then Kelly Turkeys will fetch them back, kill them and prepare them for the oven. You get paid for rearing them. You are committed to selling these turkeys, so you have them back as oven-ready birds, paying Kelly Turkeys a reasonable price for them and then selling them at a higher price. If you are killing and preparing turkeys on the farm, you must abide by current DEFRA rules and regulations.

Rearing Turkeys

A former student of mine, Richard, who has taken over his family's farm, has successfully reared turkeys for years. He does a tremendous job producing first-class turkeys. He enjoys the work and makes a profit, taking the birds through from a day old to oven-ready and keeping some indoors and some on free range.

Indoor Turkey Rearing

Two thousand turkeys are housed in a light airy building with Yorkshire board cladding. Two ends of the building are open, but in poor weather conditions, these ends can be closed with roller blinds. There are no fans, and all the lighting is natural. Many farmers, however, keep their birds in much larger windowless buildings with fans, to keep the air moving, and electric lighting. The young turkey poults are purchased at a day old during the last 2 or 3 days in July, and they are beak-trimmed before they arrive. They are purchased from Kelly Farms. The poults are given heat until they are 3–4 weeks old: 37°C under the brooder. They are fed off egg trays for the first 5 days and then will receive their feed from pan feeders. They are fed ad lib, starting off with turkey starter crumbs, then after 2 weeks the feed is changed slowly to a turkey starter pellet, and then 4 weeks later it is changed again to rearer pellets. It is not until mid-October that the turkeys receive the finisher pellets.

Turkeys aged 16 weeks with about 3 weeks left before slaughter starts.

The key factors in rearing turkeys are to avoid stress, have plenty of ventilation, ensure that they are not overcrowded and keep the litter clean. They are not vaccinated with anything but looked after with a high level of stockmanship. Killing starts on 12 December. The area where the birds are prepared is very clean and hygienic, and a real credit to the farmer.

The smallest turkey Richard produces is the Kelly Super Mini: oven-ready stags would weigh about 7.6 kg (16.75 lb), and oven-ready hens would weigh about 4.9 kg (10.80 lb). An average turkey is the Kelly Wrolstad: oven-ready stags would weigh about 9.2 kg (20.2 lb), and oven-ready hens would weigh about 5.9 kg (13 lb). His birds are not Farm Assured and not Freedom Foods, but he commands a price that is no different from those that are. Turkeys are sold privately to customers, butchers shops and restaurants, and also via his website. He told me that he killed them humanely, but I did suggest that he tried a sage-and-onion bullet, which would shoot them and stuff them at the same time.

Free Range

Well-drained fields are essential for free-range turkeys. A wet autumn and wet start to December on wet clay land is bad news for turkeys. The EU rules and regulations need checking out before you begin. The turkeys need to be brought in at night with lights on in the building, which will encourage the birds to move indoors at dusk. Electric netting will help prevent them ranging too far and will discourage foxes.

Turkeys on free range aged 16 weeks.

Free-range turkey production has certainly increased in popularity over the last few years. Turkeys can also be kept organically – check the up-to-date EU regulations and check the Soil Association rules.

Keeping Turkeys for the Hobby Farmer

You might want to keep a few turkeys, breed from them and show them. The old-fashioned breeds are hardy and easy to keep. You need a building the size of a garden shed, but if you use a garden shed you will need to improve the ventilation. You would house six in a shed 3.658 metres (12') × 2.438 metres (8'), but they would need an outside run. The wire needs to be at least 1.829 metres (6') high and the birds may even fly over this, so a covered wire run is best. The wire roof will also keep out pigeons, jackdaws and starlings, which may spread disease. Inside the shed, they will need perches to roost on. Alternatively they can go out on free range in the daytime, but they must be shut in at night because of foxes. You need to get them in before dusk, as they will fly up and roost in trees, given half a chance.

Outdoor runs or free range can get very muddy in winter, which is not a pleasant environment for the turkeys, so you may need to keep them indoors for a time, and here your ventilation should be good. Turkeys like to be dry, but they can stand the cold.

They can be quite noisy. The stags gobble, and hens make a variety of noises, so if you are going to keep them in your back garden, you had better check with your neighbours.

Turkeys mate from late January onwards. The hens are fitted with a saddle made out of leather or canvas, which the bird wears to prevent the stag damaging the hens. Eggs take 28 days to hatch. Turkeys will sit, or you can put the eggs under a broody hen or in an incubator. The young turkeys (poults) will need turkey starter crumbs, not chick crumbs, and as they grow they require grower pellets at 4–5 weeks old. If they are to be reared for the table, they will need finisher pellets at about 17 weeks old. However, if they are to be kept for breeding, they will want turkey breeder pellets plus wheat and mixed corn. They also like green food such as cabbage, Brussels sprouts and fruit such as apples. Fresh water is essential.

Turkey Breeds

Kelly Turkey Farms at Danbury in Essex have developed and supply hybrid turkeys for both indoor rearing and free range. Birds that are excellent for indoor rearing include the Kelly Super Mini and the Kelly Wrolstad. The Kelly Bronze is excellent for outdoor rearing.

Kelly Turkeys also market two other turkeys: the Roly Poly, which is a mid-size turkey between the Super Mini and the Wrolstad; and the Plumpy, which is a larger turkey than the Wrolstad.

Pure Breeds

These breeds won't provide you with a large income but are good for the hobby farmer to keep, breed and possibly show.

Bronze

Bronze originated in Europe. At one time, it was nearly extinct, but now it has come back into favour. It is slower-growing than hybrids. It had been used in developing crossbred birds, and certainly their blood has been used in the Kelly Bronze. Adult stags will weigh about 13.6–18.1 kg (30–40 lb), and adult hens will weigh about 6.4–10 kg (14–22 lb).

Bronze.

Buff (Jersey Buff)

Buff.

The original Buff turkeys became extinct by the early 20th century. However, in the 1940s, a new strain of Buff turkeys was created in New Jersey, USA. This breed of turkey has a reddish buff-coloured body with white primary and secondary wing feathers with buff shading. The tail feathers are white with a light buff bar across the rear end. They produce a good carcass and are good egg layers. They are ideal for the hobby farmer. Mature stags will weigh 10–17.7 kg (22–28 lb), and mature hens will weigh 5.4–8.1 kg (12–18 lb).

Norfolk Black

Norfolk Black.

Originating in South America, Norfolk Black are a striking jet black in colour. They are slower-growing than modern hybrids and are ideal for free range. Celebrity chefs on television sing their praises. The meat has a slight game flavour. Oven-ready stags will weigh 8.2 kg (18 lb) to 9.5 kg (21 lb), and hens will weigh 4.5 kg (10 lb) to 5.4 kg (12 lb).

Pied (Crollwitzer)

Pied (Crollwitzer).

These turkeys originate from Europe and date back to the 1700s. They are a very attractive bird with black and white markings. They are classified as an ornamental breed, but they are also good egg layers; however, they do not make a good table bird. Adult stags will weigh about 9–10 kg (20–22 lb), and adult hens will weigh about 5.4 kg (12 lb).

Other turkey breeds include the Bourbon Red, Buff, Lavender Blue, Narragansett and Slate Blue.

Chapter 20

Duck, Goose and Turkey Diseases and Ailments

Duck Diseases and Ailments

Ducks are usually healthy and do not suffer from disease very often. Like other livestock, you must have well-ventilated buildings that are not overcrowded and are kept clean. If you have a problem and are inexperienced, consult your veterinary surgeon.

Angel Wing

This is when a wing projects outwards because of too much protein in the diet. The wing has to be taped up in its natural position for at least 3 days.

Coccidiosis

This is caused by a protozoan parasite that is found in the intestines. Dirty pens with wet litter will harbour this parasite. It is not common in ducks, but symptoms include birds becoming thin and unhappy, and you may find blood in diarrhoea. Wet litter and poorly ventilated houses will harbour the disease. Birds ingest the coccidia, which, once ingested, will multiply very quickly indeed. Medication in the water such as Paracox is beneficial.

Duck Viral Enteritis (Duck Plague)

This is not a common disease in Great Britain, but there have been a few cases. It is caused by a herpes virus that can also affect geese and swans. It is contagious and attacks the vascular system. Sudden death may be the first symptom, but other symptoms are bloody nasal discharge and a bloody vent. Contact between birds will spread it, and faeces will be infected as well as contaminated equipment such as feeders, drinkers, footwear and the drinking-water. Wild ducks and geese can bring the disease in, so try to keep them out of your pens. Vaccination is available, but infected birds need culling. The main symptoms are diarrhoea that is greenish in colour, and birds become dehydrated, thin and weak. They will drop their wings; some die, and some survive but then carry the disease.

Duck Viral Hepatitis

This is a viral disease that attacks ducklings. It is a fatal disease that is contagious. Ducklings are found dead with their heads in a backward position, but you may see spasmodic contractions of the ducks' legs beforehand – they will die very quickly indeed. The liver, spleen and kidneys may become enlarged. Prevent the disease spreading by isolating the infected ducklings; there is no cure. Females used for breeding can be vaccinated against the disease.

Egg Binding

In egg binding, the duck finds it difficult to pass her egg, which becomes stuck in the oviduct. Place the vent in warm water and massage it with olive oil or lubricant, and you should be able to free it. Without help, the duck may die.

Egg Peritonitis

This is when the yolk is laid internally and deposited in the abdominal cavity. The yolk numbers build up, making the duck feel uncomfortable. The abdomen may become infected, and the duck will often die. There is no treatment.

Mycoplasma

This is a respiratory disease caused by mycoplasma that is breathed in. Birds will have a discharge from the nostrils, sneeze and have swollen eyes with foam at the corners and noisy breathing. It can be spread by contact with wellingtons, water drinkers and feeders, and wild birds. Treat with a mycoplasma-specific antibiotic – your vet will be able to give dosage details.

Respiratory Diseases

Ducks can get respiratory diseases that are often caused by houses with poor ventilation. Aspergillosis is one disease that is caused by a fungus. Fungus spores are breathed in; once the bird has the disease, it is quite difficult to get rid of it, and the birds may die. The spores will live on wet, dirty bedding. The birds will gasp for breath and lose weight. It is more common is young birds but can affect older birds. If the disease is caught early enough, it can be treated with antifungal products – however, antifungal drugs are expensive, very toxic and not that effective, so they are rarely justified in the average bird. The condition is called farmer's lung in humans.

Salmonellosis: Salmonella Enteritidis and Salmonella Typhimurium

In this condition, young birds will look unwell, have diarrhoea, become dehydrated and have ruffled feathers, and many will die. Birds become infected through bacteria in their droppings, or from a contaminated environment such as from litter, rats, mice and birds, plus dirty faecal contamination of eggs for hatching. Eggs to be hatched can be washed with Virkon in the water to kill bacteria. Various drugs will treat the disease, including neomycin and amoxicillin. Consult your vet for the best one to use and for dosage details. A vaccine is available.

Parasites

External

Mites and lice can affect ducks and geese, but they are good at preening and washing, and so the parasites are not common – I have never found any.

Internal

Gizzard worms, caecal worms, gapeworms and roundworms can all affect ducks and geese. Ducks with worms will lose weight and may have a blood-stained vent. Worms can also be found in the faeces. Broiler ducks will not need worming, but breeding stock should be wormed twice a year. Flubenvet is the only licensed poultry wormer and is a very good product indeed. Ducks can be treated with Baycox in the water; however, Baycox only affects coccidia and has no effect on worms. Check with your vet for dosage details.

Goose Diseases and Ailments

Angel Wing

The goose wing will project outwards because of too much protein in the diet. The wing needs to be taped up in its natural position for at least 3 days. If you have a problem, and you are inexperienced, consult your veterinary surgeon.

Aspergillosis

This is a respiratory disease that affects the lungs – especially in goslings. It can also affect ducks, chickens and turkeys. This disease is called farmer's lung in humans and is caused by a fungus. The spores are inhaled, and the birds will look unwell. They will gasp for breath, go off their feed and become thirsty. They will become weak, and many will die. The spores live on wet and dirty bedding. If the disease is caught early enough, it can be treated with antifungal products. However, antifungal drugs are expensive, very toxic and not that effective, so are rarely justified in the average bird. Check with your vet for the up-to-date products.

Coccidiosis

This is a disease caused by a protozoal parasite that affects mainly young geese. Keep the birds clean and avoid letting litter become dirty and wet. Birds will look unhappy and drop their wings. They will have diarrhoea, which may have blood in it. Birds ingest the coccidia, and then the parasites multiply very quickly. One form affects the kidneys and cloaca, and another form affects the intestines. Medication in the water such as Paracox is beneficial.

Duck Viral Enteritis (Duck Plague)

This is caused by a herpes virus, but it is not common in geese. It attacks the vascular system. Contact between birds will spread it, as will a contaminated environment such as through equipment. Sudden death may be the first symptom, but symptoms could include birds becoming dehydrated, having greenish diarrhoea, dropping their wings and having a bloody nasal discharge and bloody vent. Wild ducks and geese can bring the disease in, so try to keep them out of your pens. Infected birds need culling. Vaccination is available, but survivors can carry the disease.

Mycoplasma

This is a respiratory disease caused by a mycoplasma that is breathed in. Birds will have a discharge from the nostrils, sneeze and have swollen eyes with foam at the corners and noisy breathing. It can be spread by contact with wellingtons, water drinkers and feeders, and wild birds. Treat with a mycoplasma-specific antibiotic – your vet will be able to provide dosage details.

Parvovirus Disease or Derzy's Disease – Goose Hepatitis, Goose Plague, Influenza

This is a contagious viral disease that affects young geese and can also affect Muscovy ducks. They will eat less food, lose weight and have swollen eye lids, nasal discharge, diarrhoea, swollen heads and weak legs. It will kill young goslings, but mortality is very low in birds after 4–5 weeks old. There is no treatment, but a vaccine is available.

Salmonellosis: Salmonella Enteritidis and Salmonella Typhimurium
Young birds will look unwell, have ruffled up feathers and have diarrhoea, and many will die. Wild birds, rats and mice carry the disease. Birds become infected through bacteria in their droppings and from contaminated environments. Dirty faecal contamination of eggs for hatching is also a source of infection. Various drugs will treat it, including neomycin and amoxicillin. Consult your vet for the best one to use and for dosage details.

Parasites
External
Mites and lice are not common in geese because they are good at preening themselves and washing.

Internal
Gizzard worms are nematode parasitic worms. The lining of the gizzard is damaged by these red worms, and goslings can die because of them. Birds will be slow to grow and also lose condition. Try to paddock-graze your geese to give the land a rest. Treat worms with Flubenvet – contact your vet for details of dosage.

Turkey Diseases and Ailments

If you have a problem, and you are inexperienced, consult your veterinary surgeon.

Blackhead Disease or Histomaniasis
This disease is caused by a protozoan parasite carried by a *Heterakis* worm that can survive in the environment for many years – especially in the litter. Birds will eat less, have dropping feathers, and look dull and sleepy with yellow faeces. There are no drugs for treatment or prevention. Good worm control is essential – contact your vet for details.

Coccidiosis
This is caused by a protozoan parasite. The turkeys will stand hunched, close their eyes, ruffle up their feathers and have diarrhoea that is sometimes bloody. The small intestines, cloaca and rectum can all be infected. Birds living on wet litter in overcrowded houses with poor ventilation are vulnerable. Birds ingest the eggs of the parasite, which, once hatched, multiplies very quickly indeed. Medication in the water such as Paracox is beneficial.

Contact Dermatitis or Hock Burn

This is when the skin turns black on the rear of the feet, the footpad and sometimes the breast, resulting from ammonia in the litter that has burnt the skin. The black marks may develop, with lesions, ulcers and abscesses. Keep the litter clean and dry to help prevent this complaint.

Haemorrhagic Enteritis

This is a viral disease of young turkeys. Birds will feel unwell, won't feel like eating and will have diarrhoea; you may see blood at the vent, or the birds may die suddenly. The most common route of infection is oral. The death rate could be about 15%, but 60% has been seen. Rapid death may occur soon after clinical signs are observed. You can treat with oral antibiotics, but once turkeys have been sold, disinfect the building well and rest it for a month.

Mycoplasma

This is a respiratory disease caused by mycoplasma that is breathed in. Birds will have a discharge from the nostrils, sneeze and have swollen eyes with foam at the corners and noisy breathing. It can be spread by contact with wellingtons, by water drinkers and feeders, and by wild birds. Treat with a mycoplasma-specific antibiotic – your vet will be able to give dosage details.

Salmonellosis: Salmonella Enteritidis and Salmonella Typhimurium

Live poultry may have salmonella on their feathers or feet, or in their droppings. It may be present in drinkers and feeders, and spread by hands and clothing. Humans become infected when they put items contaminated with the bacteria such as fingers or other things in their mouths. Birds suffering from the disease will have diarrhoea and ruffled up feathers, will look unwell and may well die. Wild birds, rats and mice will carry the disease. Treat with various drugs including neomycin and amoxicillin – consult your vet for advice.

Parasites

Internal

Gape worms are found in the trachea but are rare in turkeys. Other species of worm are more common and are found in the intestines. The large roundworm (*Ascaridia*) is the most common. Turkeys will lose weight, be slow to grow and have diarrhoea and listlessness. Worm eggs found in the droppings will be taken in by other birds orally, and the birds will become infected. Earthworms that are eaten may carry the large roundworm, and so the turkey becomes infected in this way. Look out for worms in the droppings. It is a good idea to paddock-graze your free-range turkeys to give the land a rest.

External

If you are only keeping your turkeys for about 20 weeks, external parasites are quite rare, but if you get lice or mites, treat with an approved insecticide. However, if you keep breeding turkeys, check for external parasites, and check with your vet to see what up-to-date products are available to spray buildings and/or the birds.

Chapter 21

Arable Farming

The difference between a good arable farmer and a bad arable farmer is probably only a week! For example, the 'good farmer' may have sown his seeds before all the rain came, and then it was too wet and therefore too late for the 'bad farmer' to sow his. The 'good farmer' will have finished his harvest and all safely gathered in, but the 'bad farmer' left it a week later, and the crop was ruined by tremendous storms! If you want to know how much a crop that your neighbour has grown has yielded, don't tell him what your crop averaged first because if you do, he is sure to tell you a higher figure, e.g. if you said your wheat had averaged 10 tonnes per hectare, he is sure to say his wheat averaged 11 tonnes to the hectare!

As a first-generation farmer, you are unlikely to be an arable farmer with large acreages unless you have won the lottery or your full-time job brings you in a salary of millions. Land is very expensive, and renting a large arable farm will still take a lot of capital for equipment and machinery.

You can't make a living just growing a few hectares of wheat, barley or potatoes, and so to be a full-time arable farmer, you need hundreds of hectares. However, you can grow crops if you have a small farm owned or rented. You can have one or two arable fields. Here at Oak Tree Farm, there is a total of 13.965 hectares, and one field, 3.273 hectares, is an arable field. I purchased this field in 2009, and before purchase it had grown maize for a number of years.

You have got to decide what crop you want to grow and why. It is convenient to grow an arable crop here because if the farm was all down to grass, there would be too much grass for the number of sheep. I could keep more sheep or cattle, which would involve more man hours; I don't have to do a lot to grow a crop because contractors do the work, and the field has made a profit each year. You don't have to grow an arable crop each year in the same field; it can always be put back to grass, and you will need some sort of rotation, which may involve other fields.

Please be aware that if you farm in a Nitrate Vulnerable Zone (NVZ), you can't just put as much nitrogen fertiliser on your crops as you want; if you do, you could lose some of your payment from the Basic Payment Scheme. Amounts vary for different crops, but for example, if you are growing feed wheat, you are only allowed a maximum of 220 kg of nitrogen per hectare (although if you have historically achieved high yields, you may be able to add more). Likewise, you can only use the recommended doses for your sprays. To work out how much spray and what sprays you require, you need to have expert advice from a firm that sells the sprays. Their field officer will walk the land and tell you what you need. This is a very specialised subject indeed and needs to be constantly updated – you will not manage to grow top crops of cereals, for example, without this expert advice. My contractors, Lindley Hall Farms Ltd, employ a firm to do this, and John, from Frontier Agriculture based near Lincoln, walks my field on the same day that he walks some of theirs. He then reports back to the contractors, who then take his advice and spray. I am lucky to be in a position to have this done for me. Also note that when you buy seeds, they should have been dressed against some of the relevant pests and diseases; check when you buy them.

When considering pests in a crop, you can assess the situation by counting the number of insects on some of your plants. If you have a certain number on your crop or above that number, we say you have reached the threshold, and you need to spray. Not all crops have a threshold; for some crops, it is the percentage of your crop that is attacked instead of counting numbers.

Examples of some of the pests reaching the threshold and affecting crops include the following:

- grain aphid – five or more per ear at the flowering stage;
- gout fly – eggs on 50% of plants for winter crops;
- leatherjackets – (in spring cereals) 50 leatherjackets per square metre.

To sow crops, modern seed drills are marvellous machines, sowing the seeds very quickly indeed. My wife, Sarah, is a direct descendant of the Smyths of Peasenhall, Suffolk. James Smyth, her Great (×5) Uncle, designed and started to manufacture the seed drills in 1800. Sarah's Great (×5) Grandfather Jonathan Smyth was also based at Peasenhall until he later moved to Sweffling, where he established his own business manufacturing drills.

Smyth drills were world famous, and the drills have been described as the best horse-drawn implements ever made. It was a flourishing business, and later drills were manufactured to be towed by tractors. At its height, the Smyth seed drill business employed hundreds of workers, and it was a great asset to the village of Peasenhall. Sadly, it was sold in 1962. Houses have been built on the site of the factory, and the access

Sarah's Smyth seed drill.

road to this small estate bears the name of Smyth. Sarah now owns one Smyth seed drill, which will be passed onto our children, Jonathan and Roseanna.

Farm Assurance

Farm Assurance is product certification for agricultural produce (based on high-quality management system for food safety). Their aim is to provide consumers and retailers with confidence about product quality attributes including food safety and environmental protection. The Red Tractor Farm Assurance mark is now very well known. They set standards for farms, and you will be inspected to see if you maintain those standards and follow the rules, meeting all legal requirements.

The Red Tractor Farm Assurance Combinable Crops and Sugar Beet Scheme (combinable means harvesting with a combine and is for arable farmers). You need to check the website (http://assurance.redtractor.org.uk) for all the up-to-date information.

Wheat

Great Britain produces about 15 million tonnes of wheat a year, and about 25% of this is exported. However, this figure is dropping because of wheat being grown for biofuels.

Wheat is grown for milling and is used for bread and biscuit making as well as for cakes, breakfast cereals, pasta and some alcoholic drinks. Nowadays, it is also used in bioethanol, which is blended with petrol, and of course some wheat is grown as seed wheat for next year's crop. However, 40% of the British wheat crop is grown for animal feed, which is usually fed crushed, rolled or ground, but it can be fed to poultry whole.

Wheat can be sown in the autumn (late September to October) and is called winter wheat, and spring wheat can also be sown (February to March). The land can be ploughed and then harrowed or power-harrowed, or you can use various cultivation operations, or the quicker way of getting the land ready is to use a single-pass cultivator and then go over it with a power harrow to achieve a seed bed. The seed rate is approximately 188 kg per hectare for winter wheat, 188–251 kg per hectare for spring wheat. The days have gone when we would say 'One for the mouse, one for the crow, one to rot and one to grow!'

The crop will need fertiliser according to the state of the land, but nitrogen is the most important. The most surprising thing about growing a crop of wheat is the number of sprays that are used to get the best possible yields.

Harvesting with a combine is carried out in August and September when the wheat is ripe. Ideally, it wants to be harvested at a moisture content of 15%. Our contractors don't combine if the moisture content is over 18% unless it is a desperate situation to combine, because it takes a lot of energy to dry the grain down to 15% or preferably just under to sell it. You can combine at 20–22%, but the grain would take a great deal of drying.

Drilling winter wheat at Oak Tree Farm on 12 October 2010.

Yield has increased over the years – in 1960, the average yield would be 24 cwt per acre (3 tonnes per hectare) with a good yield achieving 30–35 cwt per acre (3.75–4.4 tonnes per hectare). Nowadays, an average yield would be about 3.12 tonnes per acre (7.8 tonnes per hectare), with good farmers on better land averaging 4 tonnes per acre (10 tonnes per hectare).

If you have only got one cereal field, it is no good going out and buying all the latest machinery to work the field. You can buy a second-hand tractor and a few implements, or you can always get contractors to do the cultivations, and you certainly need contractors to combine; perhaps you could get a second-hand baler to bale the straw or again use contractors to do this. Wheat straw is excellent for animal bedding and is baled with a conventional baler or baled into big round or square bales.

The main problem you have got with doing the work yourself is the spraying. Most farmers now use different sprays for weed and disease control; your problem is buying the small amounts of spray that you need.

I am extremely lucky in that local contractors who carry out most of the tractor work on this farm agreed to do all the work, i.e. using a single-pass cultivator, power-harrowing, drilling the seed, rolling, spreading fertiliser, spraying on chemicals, combining and baling the straw. They purchase the sprays in large quantities, usually in 5 litre cans, as they farm a large acreage, so they are in a position to mix up the sprays at their farm, come onto this farm and spray and then send the bill for the small amount of spray used. Therefore, on a small acreage, you really need contractors to do all the work. My contractors also buy my wheat for their own use. If I kept it, I would possibly have to dry it, store it and sell it. I am not 'Farm Assured,' and so many companies would not be interested in purchasing it because my storage facilities would not be good enough to meet their criteria. Alternatively, wheat can be sold 'straight off the combine' in the field, or it can go to a specialist grain store where it is stored well, and they charge you for storing it.

I could store and use the grain here, mixing it in with the poultry layers' meal. This is not a good idea, though, because the poultry ration is formulated into a balanced ration for maximum egg production, and mixing in extra wheat would 'water it down.' When you look at all the operations that are completed to grow a crop of wheat on my field, it may put you off, but as I have said before, I am lucky to have such excellent contractors; it would be very difficult indeed to do it without them.

Below is a diary of operations completed with my first crop of winter wheat on 3.273 hectares. I have listed the brand names for the chemicals – brand names differ, so you might not get the same ones as me. These branded names are chemical products, e.g. Permasect contains cypermethrin.

2009

7 September: single-pass cultivator used (not ploughed).
9 September: power-harrowed.
10 September: wheat seed sown – variety Viscount at 137.5 kg per hectare. This is a low seed rate because the crop was sown early; the later the sowing takes place, the more seed per hectare is usually required – a general rule is approximately 188 kg per hectare.
10 September: rolled.
26 October: spraying Tolugan (16.25 litres sprayed in total), an early post-emergence selective weed-killer to kill grass weed.
Permasect 0.82 litres sprayed in total, to prevent Barley Yellow Dwarf virus and Barley Yellow Mosaic virus.

2010

12 April: fertiliser spreading; nitrogen 34.5% 600 kg spread in total.
12 April: spraying Hatra 4.8 litres spread in total. A selective weed-killer, which controls wild oats, black grass and some broad-leaved weeds.
Bio power 4.0 litres spread in total. A wetting agent that makes the spray stick to the plants.
X Change 1.60 litres spread in total. This neutralises the water in the spray tank to a pH of 7.0.
23 April: fertiliser spreading nitrogen 34.5% 915 kg of product spread in total.
26 April: spraying Capalo 4 litres sprayed in total. A fungicide.
Liquid manganese 8 litres sprayed in total. To stop manganese deficiency.
Canopy 1.6 litres sprayed in total. A growth regulator that makes sure the stems (straw) are short, thus helping to stop the crop falling over.
Stabilan 4 litres sprayed in total. Another growth regulator.
Biplay 140.00 grams only. A broad-leaved weed-killer.
Joules 4 litres sprayed in total. A fungicide.
6 May: fertiliser spreading 34.5% nitrogen 470 kg of product spread in total.
26 May: spraying Prosaro 2.5 litres sprayed in total. A fungicide.
Alpha Fenpropidin 5 litres sprayed in total. A fungicide.
24 August: combined – yield 8.98 tonnes per hectare (3.59 tonnes per acre), which is above the national average. Sold for feed wheat.
24 August: straw baled and carted 42 big square bales. Some sold and some kept for bedding.

Oats

During the 1970s and 1980s, I purchased rolled oats and mixed them with other ingredients to make up a ration for my sheep. Nowadays, to make life easier for myself, I purchase a balanced sheep ration in the form of pellets, but I do think oats are a good feed for sheep.

As well as livestock feed, oats are grown for human consumption and are used in breakfast cereals and 'oaty bars.' Oats can be grown on all classes of soils but are grown more in the moist areas in the north and west. They will tolerate a more acid soil than wheat and barley.

Oats can be sown in the autumn (September/October) and spring (February/March). The land can be conventionally ploughed and then various cultivation operations completed to obtain a tilth, or it can be worked with a single-pass cultivator and then gone over with a power harrow. The seed rate is approximately 200 kg per hectare. The crop will need some fertiliser – the amount needed will vary according to the condition of the land. The crop will also need spraying.

Harvesting with a combine is carried out in August and September when the oats are ripe. Ideally it wants to be harvested with a moisture content of 15% or just under. Too much moisture means the crop has got to be dried. It usually needs to be 15% before you can sell it. Our contractors wouldn't combine over 18% unless they were desperate to combine. Yield has increased over the years, and now the average yield is about 5.8 tonnes per hectare.

Oat straw is my favourite straw and is an excellent straw for feeding to cattle or can be used as bedding. It can be baled with a conventional baler or baled into big round or square bales.

Harvesting oats at Oak Tree Farm 11 August 2012.

As I have already stated in the earlier section on wheat, it is no good going out and buying new machinery just to grow a small acreage. You can use some second-hand equipment alongside contractors, or just use contractors. You really need contractors to do your spraying because they hopefully can sell you the very small amounts of each spray that you need.

I grew oats on the 3.273 hectare arable field, sowing them in September 2011 and harvesting them in August 2012. All the work was completed by the contractors, the same ones who carried out the work for my wheat, and all the oats were purchased by this firm as well, giving me a very fair price.

I did think about keeping some back to feed to the sheep, but by the time I mixed it up with sheep pellets, I thought I might as well cash them in.

Below is a diary of operations completed with my first crop of winter oats (3.273 hectares). I have listed the brand names for the chemicals; brand names differ, so you might not get the same ones as me. These branded names are chemical products, e.g. Tolugan contains chlorotoluron.

2011

29 August: single-pass cultivator used (not ploughed; the oat crop will follow two crops of wheat).

9 September: spraying – Clinic Ace 6.5 litres in total. A non-selective weed-killer plus X Change 1 litre spayed in total. This neutralises the water in the spray tank to a pH of 7.0 plus Blaze 3.5 litres sprayed in total. Makes the spray stick to the plants. Spraying before further cultivation produces a stale seed bed killing all the weeds.

20 September: power-harrowed.

20 September: winter oats drilled variety Gerald at the rate of 152 kg per hectare.

18 October: spaying Lexus Class 195 grams sprayed in total. A selective weed-killer that kills some grass weeds and some broad-leaved weeds.

Permasect 0.81 litres sprayed in total to prevent Barley Yellow Dwarf virus and Barley Yellow Mosaic virus.

2012

30 March: spraying Eagle 81 grams spread in total. A broad-leaved weed-killer.

Optimus 1.3 litres sprayed in total. A plant-growth regulator – makes sure the straw will be short, thus helping to stop the crop falling over.

Capalo 1.65 litres sprayed in total. A fungicide.

Liquid manganese 6.5 litres sprayed in total. To stop manganese deficiency.

8 April: fertiliser spreading – Nitraprill 600 kg spread in total.

10 August: combining. The crop yielded 7.75 tonnes per hectare, which I was very pleased

with; this is well above the national average.

11 August: baling – 39 big bales – sold.

14 August: conventional baled 205 bales – carted and stacked – kept for bedding.

Barley

This is the second most important cereal crop after wheat. Barley is used for malting, for beer and whisky making, and is also used in health foods, soups, stews and drinks; about 50% of the British crop is grown for animal feed. It does not yield as much per hectare as wheat.

Malting barley needs to be of very good quality, and feeding barley is not so specialised, so for the first-time farmer, I would start with feeding barley.

Barley will grow in poorer conditions than wheat and can be sown in the autumn – end of September/October and November. Spring barley is sown in February/March. Winter barley is higher-yielding than spring barley, but it is usually the spring varieties that are used for malting, and if the crop makes the grade for malting, it should command a higher price than feed barley.

Over the years, I have fed tonnes of rolled barley to cattle and fed milled barley in pig rations, but I prefer to feed oats to sheep. If you do decide to grow a field of barley for animal feed, make sure you can sell it, or if you have got storage facilities, you can keep it on the farm and use it. As I have explained in the wheat and oats sections, you really need contractors to do all or most of the work.

To grow a crop of barley, you don't go by a rule book 100% because it will depend on your soil type, what nutrients are in the soil and the climate. If you are growing winter barley, and it is a very wet autumn, you may be late sowing the crop because it is not fit to work the land. Tractor work may easily vary by a month from year to year or even on different fields in the same year. In some years through a lack of sun, the crop may be slow to ripen, and in some years you will need more sprays owing to more weed growth or disease.

If I was going to grow a crop of winter barley, I would more or less carry out the following procedures. During the middle of September, the field may need a non-selective weed-killer spraying on it before starting cultivations – 'Roundup' or a similar product. Depending on the previous crop, the field will want ploughing and cultivating, or use a single-pass cultivator and then probably power harrowing; barley needs a fine seed bed. Sowing will take place towards the end of September using a feed variety such as Cassata, sowing it at a rate of about 160 kg per hectare. The field is then rolled. As soon after sowing as possible, a pre-emergence weed-killer would be applied to kill grass weeds and broad-leaved weeds. During the middle of October, it would be sprayed to prevent Barley Yellow Dwarf virus and Barley Yellow Mosaic virus. The field would then be left until the spring.

Barley almost ready for harvest July 2013.

At the end of March, it would benefit from some nitrogen fertiliser – nitrogen 34.5%. How much would depend on what the field had growing on it last year, but about 250 kg per hectare would be about average. During April, the field would be sprayed with a selective weed-killer to kill broad-leaved weeds and also sprayed with a fungicide. In early May, more nitrogen would be applied, 34.5% nitrogen, 250 kg per hectare, and also a fungicide.

Harvest when ripe in July – the average yield in Britain is 5.7 tonnes per hectare. Barley straw is good for feeding cattle, but I prefer silage, especially for dairy cattle.

Potato Growing

Potatoes are the third most important food crop after rice and wheat. In Britain, approximately 80% of the crop is for human consumption, including chips and crisps; the rest goes for seed potatoes and livestock feed.

You might want to grow a few potatoes to sell at the farm gate, and if you only want a small amount, this can be done with a small amount of machinery or indeed a combination of machinery and hard work. After leaving school, I worked for a year on a local farm before attending college. One of the first jobs I was asked to do was to dig up the potatoes in the farm garden. I imagined this to be an easy job taking me half a day to complete at the most. However, I was taken into the garden, which looked like a small field and it took 10 days to dig them up with a garden fork, and then I had to bag them – but you can do this if you want to grow them on a small scale.

If you decide to grow a field of potatoes, you will need to organise contractors to carry out the work. Potatoes are an expensive crop to produce. The field needs careful preparation, you need to buy the seed potatoes plus fertilisers and sprays, and then you need to harvest the crop. Spraying for potato blight is important, and potatoes may need spraying every 10 days throughout the growing season.

You need to make sure you can sell all your potatoes before you start. If you are selling some at the farm gate, you need to think about how you are going to store them before they are sold – in boxes in a controlled environment is best, but this is expensive. All your potatoes can be sold wholesale, which saves you the bother, but of course you will get less money than if you sell them at the farm gate.

A number of years ago, one of my friends, Myles, grew 'pick your own potatoes.' This involves growing a field of potatoes as normal; then at harvest time, the public come in and pick them off the ground. The event is advertised, a grass field is made available for a car park, and you or a trustworthy employee need to take the money. A tractor and an old-fashioned potato spinner spins the potatoes out of the ground. The customers are given a bag, and they pick up what they want. The full bags are weighed in the field and are paid for. At the same time, Myles's son, Richard, was in my class at school. Richard had been absent for quite some time, and the headmaster sent the truant officer or 'Wag man' to the farm. He found Richard helping in the field with the potatoes. Richard did return to school the next day. Soon after this, I met Myles and said 'I hear you had a visit from the truant officer.' 'Yes, John,' he replied, 'I gave him a bag of potatoes, and so it was all soon sorted out.' When the headmaster found out he said that he would visit the farm next time!

You have to decide when you stay open, and you need a spell of fine weather, as the public will not be keen to pick the potatoes in a muddy field on a wet day, and they will worry about parking their cars on grass. You will need public-liability insurance in case of accidents. If you farm 'pick your own strawberries,' customers will eat some in the field without paying. With 'pick your own potatoes,' customers will not eat them in the field!

You need to decide what to do with any potatoes that are left; they will need to be lifted before winter and then sold or stored.

To grow potatoes, you need a soil that can be worked to a good depth that is not too stoney. Potatoes like plenty of organic matter in the soil and plenty of fertiliser, and they need to be sprayed on a regular basis for potato blight and sometimes for aphids. They are prone to pests and diseases, and so are often grown as part of a 5-year rotation.

Potatoes are classified into first earlies, second earlies and main crop. First earlies are planted as early as possible in the year. They are mainly grown in areas not prone to frosts such as South Wales and the Channel Islands. These small potatoes are sold in the shops at the end of May. They taste good and usually command a good price. Second earlies and main crop are the crops that are mostly grown in Britain. Typical yield in Great Britain is about 45 tonnes per hectare. Seed potatoes are grown in the north on higher ground where aphids are uncommon – aphids spread viral diseases, which are bad news for potatoes.

My near neighbours, John, Richard and Robert, are noted for their excellent potato growing. They grow about 30 hectares each year, selling as many as possible in the farm shop and then selling to other farm shops, and fish and chip shops, and what is left is sold wholesale.

A number of varieties are grown, but Wilja, Maris Piper and King Edward would be at the forefront. A few first earlies are also grown, Maris Bard being a good variety that has white skin and tastes like 'new potato.' Frost can be a problem, so most of the crop is second earlies and main crop.

The land is sprayed with a non-selective weed-killer such as 'Roundup.' It is then ploughed and a tilth created with a single-pass cultivator.

Seed potatoes are purchased from Scotland, the chits (shoots) will have just started to grow. The first earlies are not planted too early – the first week in April is about average – too early may mean frost damage. Most of the potatoes are planted later on during April at the rate of approximately 70 kg per hectare. The rows are 91.5 cm apart (36 inches); Wilja is a second early and is planted 35.5 cm (14 inches) between each potato in the row, and King Edward is a main crop and is planted 45.7 cm (18 inches) between each potato in the row. A machine moves the stones and clods to one side; fertiliser is added at the rate of 150–200 kg of nitrogen, 170 kg of phosphate and 300 kg of potash per hectare, half of which is spread on the top of the soil and half sown with the potatoes actually in the ground. Just before the potatoes come up, a pre-emergence weed-killer is applied to kill broad-leaved weeds and grass weeds. To give the crop a boost, and if it is felt it is needed, an extra application of nitrogen is added at the rate of only about 30 kg per hectare, and this is spread when the leaves are just starting to meet in the rows.

Spraying will have been started before this – soon after potato growth appears – when the plants are palm-sized.

Depending on the weather – the ideal is to spray against potato blight every 10 days, and not really leaving it longer than 14 days if possible. This spraying goes on and on right up to when the crop is sprayed off to kill the tops with a contact herbicide called Diquat. Diquat is a desiccant, which means it causes the tops to dry out quickly and die. The tops would die off naturally, but a few roots of potatoes are dug up, and if they are the right size for harvesting, Diquat is used. Because the tops die, this has the effect of setting the skins on the potatoes, which helps storage.

As well as spraying probably at least six times for potato blight during the growing period, the crop may want spraying for aphids; the crop is examined on a very regular basis, and if an average of five aphids appear on each plant, the crop is sprayed with aphicide. Potatoes are harvested from the middle of September onwards. They are taken back to the farm, sorted and then sold in bags in the shop. Potatoes for storage are stored in potato boxes in an environmentally controlled building that is insulated and thermostatically controlled, keeping the potatoes at approximately 5°C.

An average yield for the farm is 55 tonnes per hectare. The above is a very successful enterprise and a credit to the farmers. Farmers will have weather warnings on potato blight called 'Smiths periods.' When the minimum temperature is above 10°C and humidity above 90% for 11 hours a day for two consecutive days, this is a 'Smiths period,' and the crop is at risk, so you would probably spray. Big potato growers will have their own weather station to monitor the blight period themselves.

Potatoes being harvested 22 September 2013.

Root Crops

Root crops include turnips, swedes, mangolds, sugar beet, fodder beet and carrots. Carrots are grown in market gardens or on sandy soils and need good growing conditions. Mangolds seem to be out of fashion – certainly in this area. If you want to grow a root crop, I would suggest turnips, swedes or fodder beet.

Stubble Turnips

Turnips are a good old-fashioned crop that is very easy to grow and can be fed to cattle and sheep. They are particularly beneficial for lambs to graze in the autumn and winter. They are good to include in a rotation, giving the land a break from cereals and grass. I grow just 1011.7 square metres of turnips at the rear of the poultry building. The land is churned up each year when the poultry are cleaned out, and the area often becomes a muddy mess. It is not worth sowing grass seed on it for it to be churned up the next year, and so turnips are ideal, as they are ready to graze after 12 weeks. They can be sown in April, May or June for summer grazing, but they can also be sown in September, and then this crop will feed the sheep in early winter when the grass is scarce. Many farmers sow them after a cereal crop, and these are called stubble turnips. You can sow them directly onto the stubble, but a fine seed bed is best because the seeds are so small. The land is often worked with a single-pass cultivator and then possibly a power harrow and a roller. You can sow the seed in rows with a precision drill or broadcast. I sow them broadcast. You only sow 5 kg per hectare, 1 cm deep. This is a very small amount of seed indeed, and when you look at it, you can't believe it will be enough. The seeds are very small, and if you sow the seed too thickly, the plants won't have enough room to develop a decent-sized root, so you will end up with a lot of top growth but very little root formation. A pre-emergence weed-killer may be applied to kill broad-leaved weeds and grass weeds. Turnips are one of the quickest seeds to germinate, and it is amazing how quickly the field is covered with plants. You can give them some fertiliser in the seed bed. The land may want analysing to see if it requires fertiliser – turnips are a quick-growing crop and may need very little – especially if it has had plenty of manure: 75 kg of nitrogen, 40 kg of phosphate and 40 kg of potash per hectare is plenty, and a top dressing of nitrogen about 75 kg per hectare a month after sowing can be

Store lambs eating stubble turnips 30 January 2014.

applied. I apply no fertiliser to my crop, as the ground it is growing on is rich in poultry manure. A pH of 6.5 is good, but my land for turnips is more acidic. Strip grazing of the crop is best using an electric fence; this minimises wastage.

You can also purchase main-crop turnips that take longer to grow – about 20 weeks. They are hardy, and you can graze them well into winter. Brassica crops contain some toxic substances including nitrates, sulphur and molybdenum, so be careful – introduce sheep gradually and leave them on for a while, then bring them back onto grass. You could upset their digestive system and could even have some die. An average fresh yield would be 38–40 tonnes per hectare for stubble turnips.

Swedes

Swedes are another good old-fashioned crop and can be grown for cattle or sheep. They are usually grazed in the field. An advantage of swedes is that they will grow in the wetter, cooler parts in the north and west.

A good fine, firm seed bed is required, and so you can either plough and cultivate or use a single-pass cultivator with possibly a power harrow and roller.

The seeds are very small, and the seed rate will vary depending on how the seed is drilled or broadcast. Check with your seed merchant on how much to sow per hectare, but it will be approximately 3.75 kg per hectare at a depth of only 1–2 cm. Sow treated seed to minimise risk from pests. Sowing can take place in April, May or June. I prefer April or early May, as it could come very dry in late May or June. Have your soil analysed to see what fertiliser is required. Some nitrogen, phosphorus and potash is best applied in the seed bed, which could be about 250 kg in total per hectare, depending on what the soil needs and what crop was grown previously. Swedes don't like acidic conditions, and a pH of 6.5 is ideal. A pre-emergence weed-killer may be applied to kill broad-leaved weeds and grass weeds.

Swedes fed to Derbyshire Gritstones.

Swedes are ready to be grazed in the autumn and winter. Watch out for mildew, which reduces palatability and yield, and also watch out for club root. The average fresh yield is about 75 tonnes per hectare.

Fodder Beet

Fodder beet is a good root crop to grow and can be fed to cattle or sheep. It is harvested and brought to them rather than grazed in the field.

Ray, one of my near neighbours, keeps a lot of sheep and has grown fodder beet for years. In the past, I have bought fodder beet from him to feed my sheep. He grows the crop as follows. The land is always ploughed – it is a good deep soil; after ploughing it is cultivated, and a fine clean seed bed is created. Drilling takes place from the end of March until mid-May. The seed is pelleted and sold conveniently in 1 acre packs, each pack containing 50,000 seeds. The seed is drilled in rows at a depth of 2.5–3 cm.

Fertiliser may be added to the seed bed – no nitrogen, some phosphorus and mainly potash, but with so many sheep on his farm, the land is very rich in nutrients and organic matter, and so often no fertiliser is needed. (I would suggest a soil analysis – fodder beet crops benefit from sodium and magnesium and also the trace elements boron and manganese; apply these before the seed bed is prepared – sodium, for example, could damage your seedlings if spread too close to drilling.)

A pre-emergence weed-killer is sprayed on the land, which kills broad-leaved weeds and grass weeds. Broad-leaved weeds and grass weeds are then sprayed again when the crop is at the second leaf stage and again a fortnight later; 'fat hen' is one of the main weeds.

Fodder beet growing well 19 September 2013.

The crop is checked regularly for flea beetle, and if it is found it is sprayed with insecticide. The crop is harvested with a machine during October, giving a fresh yield of 50–90 tonnes per hectare. Some farms can achieve 100 tonnes per hectare. Sheep may be let into the field to clean up the tops after they have wilted – the tops may yield 10–20 tonnes per hectare.

Forage Crops

Forage crops are crops that provide green food for livestock that can be grazed in the field or 'zero-grazed,' i.e. brought in and fed to the livestock. Forage crops include brassicas, such as cabbage, kale, mustard and forage rape; leguminous crops, such as lupins and vetches; and cereal crops, such as maize and rye.

Probably the best crops to grow are forage rape, kale and maize. Maize will be dealt with in Chapter 23, 'Haymaking and Silage Making.' However, be careful because brassica crops contain toxic substances including nitrates, sulphur and molybdenum. Brassica crops can also prevent thyroxin production, which can cause goitre. These crops must not be over eaten.

You can also purchase a mixture of seeds of both root crops and forage crops, such as stubble turnips, forage rape and kale. The mixture is all sown together and then when ready is grazed by stock.

Forage Rape

Forage rape is similar to turnips and swedes, but it lacks the swollen root. It is good for winter feeding. The small seeds of forage rape make it sensible to prepare a fine seed bed by either ploughing and harrowing or a single-pass cultivator and possibly power-harrowing and rolling. The seed can be either drilled or broadcast. You sow less seed per acre if you drill it 6 kg per hectare, but if you broadcast you would need 10 kg per hectare; sow the seeds from April to August.

A pH of 6.5 is ideal. I would suggest a top dressing of nitrogen after a month of it being sown, about 75 kg per hectare. It is ready for livestock to eat at about 13 weeks. It would need to be strip-grazed for cattle and sheep to avoid wastage. The average fresh yield is about 30 tonnes per hectare.

Forage rape 11 November 2014.

It is a simple crop to grow, but on balance I would prefer to grow turnips, swedes or fodder beet because they have a swollen root as well as the tops, and so there is more for the stock to eat.

Kale

The growing of kale has almost disappeared from this area. When I was a lad in the 1960s and 1970s, it was a fairly common crop. I have seen it grown as game cover for pheasants and partridges, but it is rarely seen as a forage crop.

I would sooner grow kale than forage rape, as it produces a much higher yield and makes good cattle feed, which lasts longer than forage rape. However, you may need a lot of cattle to eat it. It is hardly worth growing for just a few animals.

Kale is best grown on free-draining soils; cattle grazing kale in the winter on heavy clay land can make a great deal of sludgy mess. Slugs can be a problem if the field was grass the year before – you may need to use slug pellets.

Kale is usually sown between April and mid June but can be sown up to the first week in July, albeit this may give a lighter crop. A firm, fine seed bed is essential and is obtained by ploughing and harrowing or using a single-pass cultivator with possibly a power harrow. This may need rolling before the seed is sown, but I would certainly roll after sowing to make the seed bed firm. You use less seed if you drill the crop – 5 kg per hectare, but broadcast you would sow 7.5 kg per hectare.

Kale likes organic matter and does well with good old farmyard manure, but if you are short of manure, nitrogen can be placed in the seed bed up to about 100 kg per hectare. Analysing the soil would confirm if this nitrogen is required. You may need to top dress with more nitrogen 6 weeks later.

Kale 10 November 2014.

Kale takes about 20 weeks to grow into a full crop. To control weeds, apply a pre-emergence spray straight after sowing, or alternatively you can produce a stale seed bed – that is, after ploughing and cultivation, let the weeds come up, and after about 3 weeks, spray with 'Roundup'; then soon after, sow your kale seeds.

Kale can be either strip-grazed or zero-grazed (cut and carted in to the livestock), and grazing takes place throughout the winter. The leaf is the most valuable part of the plant and is where most of the protein is found, so the crop should not be wasted. When using an electric fence, move the fence to allow cattle to graze under the wire so the crop is not walked on. The size of the field and the number of cattle will dictate how often you move the fence – once a day is often very practical.

The average fresh yield is about 60–65 tonnes per hectare. Watch out for flea beetle, slugs and club root. Some varieties are club-root-resistant; you will need one of these varieties if you grow kale in the same field for the second year. Watch out also for rabbit and pigeon damage.

Other Crops

These would include beans, peas, oil seed rape and linseed. Beans are easy to grow; peas don't like heavier soils and waterlogging, and yields are variable. You are best growing oil seed rape with contractors' help. It can be a profitable crop. Linseed is best grown under contract. I would suggest field beans and oil seed rape – both crops are harvested for their seeds.

Oil Seed Rape

Oil seed rape is harvested for its seeds. It is a cash crop and is not grazed by livestock. It is a member of the mustard and cabbage family.

When I was a boy back in the 1950s and 1960s, it was almost unheard of, but it started to become more popular in the 1970s and has now become a commonly grown crop that can be profitable. Oil seed rape fits in well in a rotation and is a good break crop from cereals. It is not grown in the same field too often, preferably at least only one year in three, because of the risk of disease. The seeds are used for a cooking oil called vegetable oil, which is high in quality, and the seeds are also used for biodiesel. There are winter varieties sown in the autumn and spring varieties.

I would grow the crop as follows. Plough and cultivate the land or use a single-pass cultivator and probably power-harrow to create a fine tilth. Drill the black seeds, which are similar to poppy seeds, at the end of August and September at 5 kg per hectare then roll. The seed will have been dressed with an insecticide and fungicide, which will give it a good start. Soon after drilling, apply a pre-emergence herbicide for broad-leaved weeds. Spray the crop in mid-October for grass weeds and spray at the end of October with an insecticide and fungicide. Leave the crop until early spring, when some fertiliser should be added – the amount will depend on the state of your soil and the previous crop – I suggest getting your soil analysed.

Oil seed rape in flower April 2014.

Apply fungicide and probably a herbicide in April to kill weeds such as mayweed. Apply the bulk of nitrogen fertiliser in late April or early May to give a maximum of 220 kg per hectare – most farms would apply about 180 kg per hectare. Sulphur may also be needed, and if pollen beetle is present it will need to be sprayed with insecticide. Apply a further fungicide at the end of flowering in late May; you may have to spray insecticide for seed weevil and pod midge.

Desiccate the crop (to kill the foliage) with 'Roundup' 10 days before harvest to make harvesting easier and more effective. This would probably be done towards the middle or end of July, depending on the season. Winter rape would yield an average of about 4.25 tonnes per hectare.

Field Beans

Field beans are a leguminous crop and are in the same family as peas and lupins. Legumes have symbiotic nitrogen-fixing bacteria in the root nodules. Field beans are a good crop to grow for the first-time farmer. They make a good break crop from cereals and are not expensive to grow, requiring little fertiliser and pesticide compared with cereals. They are high in protein (twice the protein found in cereals) and so make a good animal feed. They can be sown in the autumn (winter beans) or in the spring.

Field beans ready for harvest in September 2013.

I would grow a crop of winter beans using the following guidelines. The land should be well drained and not acidic; winter beans can be damaged by frost. A seed bed similar to that for cereals is ideal. The seeds can be drilled or broadcast and sown in October through to November. A pre-emergent herbicide should be applied soon after sowing to control some broad-leaved weeds and grass weed. Apply a grass weed-killer in April and also a fungicide. A further application of a fungicide may be needed in May plus an insecticide. Low inputs of fertiliser are used on this crop – no nitrogen, but after soil analysis you may need some phosphate and potash.

Desiccate with 'Reglone,' which dries out and kills the stems and leaves, about 2 weeks before harvest. Harvest with a combine – the average yield is about 3.75 tonnes per hectare. Winter beans should yield more than spring beans if they have not been damaged by chocolate spot, pigeons or frost.

Organic Arable Farming

When growing crops organically, you will get lower yields for your crops but hopefully more money for the finished crop. With lower yields, there would be less surplus; however, if all arable farming was organic there would not be enough land available in Great Britain to feed the nation. Environmentally, it may be a good thing because of the non-use of chemical sprays and inorganic fertilisers, and less tractor diesel would be used. Many people believe organic food is better for you, but many can't afford to pay the higher prices in the shops.

For the first-time farmer, I would suggest going non-organic to start, and then in a few years' time you might want to change to organic, but there is a conversion period – you can't go organic overnight. You will need to check with DEFRA for all the rules and regulations (and there are lots of them), and above all you should make sure you can sell all your crop as organic. As a family, we don't buy organic foods, and this farm is not organic, but this is our choice.

A woman telephoned her husband at work and asked him to buy some organic vegetables on his way home. The husband arrived at the shop and began to look for organic vegetables. He couldn't find any, so he finally asked a shop assistant where they were. The assistant didn't know what he was talking about so the husband said, 'These vegetables are for my wife – have they been sprayed with poisonous chemicals?' The assistant replied, 'No – you will have to do that yourself!'

Biofuels

Biofuels are produced from living organisms and are an alternative to fossil fuels. Biodiesel is made from vegetable oils and animal fats. Oil seed rape is a good crop to grow for biodiesel. Biodiesel can be used in its pure form but is usually used just as an additive to diesel to reduce levels of hydrocarbons and carbon monoxide.

Bioethanol is an alcohol that is made by fermentation from sugar into starch. Sugar cane, sugar beet, cereals and potatoes can be grown for bioethanol. They can be mixed with petrol to reduce carbon dioxide. Biomass is for electricity and heat. Crops grown for this include trees, coppice and giant grasses such as *Miscanthus* – elephant grass.

This all means an alternative market for some farmers. There are supporters and opponents of biofuels, and the European Parliament has backed proposals to limit the amount of food crops used to produce biofuels. The Vivergo plant near Hull will now take 1.1 million tonnes of wheat a year to turn into ethanol, which is estimated to produce 420 million litres of ethanol a year. This is all very interesting, but as with most first-generation farmers, most of the crops you grow will be for animal feed.

Genetic Modification (GM)

Genetic modification is biotechnology that is being used to modify crop plants. No GM crops are being grown commercially in the UK at the moment, but experimental trials are being carried out. Some GM crops are imported into the UK, especially soya and maize, and are being used mainly for animal feed and in some food products.

The bank manager visited my neighbour's arable farm. 'How much money do you pay your workers?' he asked. 'Well,' replied the farmer, 'David has been employed for 10 years, he works 40 hours a week, and I pay him £380.00. Alf has been employed for five years, he works 30 hours a week, and I pay him £240.00. Then there is this complete idiot who works 110 hours a week, and he takes home £20 a week.' 'That's the worker I want to discuss things with,' replied the bank manager. The farmer replied 'That's me. I'm that complete idiot!'

Chapter 22

Diseases and Pests of Arable Crops

Main Cereal Diseases

Cereals are prone to a number of diseases, but if the seed is dressed with fungicides, and you are prepared to spray your crops, you should have little trouble. These days, some cereal varieties are more resistant to certain diseases than other varieties. If your cereals have got a viral disease, it may have been transmitted by aphids; you can spray your cereals to kill the aphids, but the damage has probably been done by the virus, and it is hard to actually kill the virus. If your cereal crop has a fungus disease, you can spray with a fungicide, but this depends on a number of factors including what stage the crop is at and the state of the weather. Expert help on your spraying programme is essential in my opinion.

Name of disease	Crop attacked	Caused by	Signs of the disease
Barley Yellow Dwarf Virus	Wheat, barley, oats, maize and some grasses	A virus transmitted by aphids	Bright yellow upper leaves, plants are stunted with less yield because no heads or fewer heads are formed
Barley Yellow Mosaic Virus	All cereals – each cereal has a different form that is specific to that cereal	A virus transmitted by a soil-borne vector	Plants can be stunted, and the leaves have pale yellow streaks, which could change to purple or brown at the tip of the leaf. Later flecks of brown may replace the yellow streaking in the leaves.
Brown rust	Wheat, barley and rye	Fungus	Orange and brown pustules seen on the leaves and sometimes on the stem and glumes during the autumn
Eyespot	Wheat, barley and oats	Fungus	Brown bordered spots are shaped like eyes, which are found at the base of the plant stems
Leaf stripe	Barley	Fungus	Spots or bold long brown stripes on the leaves can result in no grain in the ears

Name of disease	Crop attacked	Caused by	Signs of the disease
Loose smut	Wheat, barley, oats – distinct forms attack the different crops	Fungus	Black smutty ears caused by fungal spores that replace the grain
Net blotch	Barley	Fungus	Brown stripes and blotches and dark lines on leaves. The blotch looks like a net, hence the name.
Powdery mildew	Wheat, barley, oats and most grasses	Fungus	Found mainly on the leaves, white pustules produce spores that are powdery; then black spore cases will occur in the mildew pustules
Rhynchosporium (leaf scald)	Barley, rye and some grasses	Fungus	Scald lesions oval in shape on ears and leaves that are green with a brown edge
Septoria	Wheat, sometimes rye	Fungus	Water-soaked patches on many leaves that turn brown and may contain black fruiting bodies
Stinking smut or bunt	Wheat	Fungus	Black spores on grain that smell like fish. Rare in Britain because seed is treated with fungicides.
Take-all 'whiteheads'	Wheat, barley and some grasses	Fungus	Black colour at the base of the stems; bleached seed heads that give rise to the name 'whiteheads.' When wheat is grown continuously year after year, in the first growing year, you would probably get no take-all; in the second and third year, you would probably get take-all; however in the fourth and fifth year there would be less risk of the disease.

Name of disease	Crop attacked	Caused by	Signs of the disease
Yellow rust	Wheat, barley and some grasses	Fungus	Lemon yellow pustules found in parallel lines on stems and leaves

Main Pests of Cereals

Insects are killed with insecticides, but you are better to prevent them attacking if you can, rather than spraying them when damage is being done. Dressed seed and varieties that are less vulnerable or more resistant to attack will certainly help, but before spraying consult an expert.

Pest	Crop attacked	Caused by	Signs of pest
Aphids	All crops	Insect	Indirect damage – aphids feed from the cereal and spread the Barley Yellow Dwarf Virus. Direct damage – aphids feeding from the leaves and stems.
Frit fly	Wheat, barley, oats	Larvae	Larvae tunnel into the shoots. Plants become stunted with fewer tillers.
Gout fly	Wheat and barley	Larvae	Found in the tillering stage of spring cereals. Stems are swollen and stunted (hence the name gout) and produce no ears of corn. Cereals sown in the autumn may still produce grain. Leaf tips have a ragged look about them and appear yellow.
Leather-jackets	All crops	Larvae	Leatherjackets are the larval stage of the cranefly or daddy long legs. The leather jackets feed underground or at the surface of the soil, often cutting off the shoot below ground level. Plants become yellow and wilted. Spring cereals grown after grass are at greatest risk of attack.
Orange wheat blossom midge and yellow wheat blossom midge	Wheat, barley, and rye but mainly wheat and rye	Larvae	Orange – damage to developing grain causing cracking of the seed-coat. Yellow – the larvae feed on the flower preventing pollination. Certain varieties are now resistant to orange blossom midge.

Pest	Crop attacked	Caused by	Signs of pest
Slug	Mainly winter cereals	Nymph and adult	Slugs will eat the new shoots, as the plant germinates, and leaves on the growing crop, resulting in complete loss of leaf. Slugs prefer heavier soils.
Wheat bulb fly	Wheat, barley and rye	Larvae	Larvae can be seen burrowing into young seedlings below ground level feeding off the central shoot, causing yellowing and loss of tillers and bare patches in the crop
Wireworm	All crops	Larvae	Wireworms are the larval stage of the click beetle. They feed on the seed and damage roots and stems resulting in crop loss.

Main Potato Diseases

Consult expert help on potato diseases that are likely to occur in your area and form a plan of preventative measures and what to do if your crop is attacked.

Name of disease	Caused by	Signs of the disease
Potato blight	Fungus	The crop is attacked under warm damp conditions. Dark blotches on the leaves appear that are often surrounded by a pale halo. These blotches spread and rot the leaves and stems. Spores are also found on the underside of the leaf, and under damp conditions the stems and leaves are reduced to a rotting mass within a few days. Dark markings are also found on the surface of the potato tubers with chestnut-coloured rot under the skin. The skin will rot away, and this may be speeded up with fungal and bacterial infection.
Potato Leaf Roll Virus	Virus transmitted by aphids	Leaves curl up or roll and appear leathery. The potato foliage is yellow and stunted; older leaves may turn brown, stunting the plant.
Scab (common and powdery)	Filamentous bacteria and fungus	Common scab – scurfy brown patches on the tubers. Powdery scab – a mass of brown spores on the surface of the potatoes.

Main Pests of Potatoes

Consult expert help on preventative measures including spraying and what to do if your crop is attacked.

Name of pest	Caused by	Signs of pest
Aphids	Aphids	Aphids may overwinter on the cabbage family and then move to the potatoes in the spring. They can be seen feeding on leaves and stems and also spread viral diseases.
Potato eelworm	Small thread-like organisms only visible under a microscope	The pest lives inside the root of the plant. The plant is affected by yellowing of the stems and leaves, and die-back of the foliage, and a poor yield may result. The eelworms puncture the plant cells and destroy tissue inside the plant. Egg-bearing cysts can be seen with a magnifying glass on the roots when the potatoes are lifted. These cysts can live in the soil for up to 10 years.

Main Diseases of Root Crops

There are many more diseases than the ones I have covered, and you will need an expert in many cases to help you identify your problem. If you have a viral disease, the damage may already have been done, and there is probably little you can do. Fungicides kill fungus, but again you will need expert advice to decide when to spray and what product to spray with. Certain varieties of crops will show resistance to some of the diseases, and dressed seed will be beneficial.

Name of disease	Crop attacked	Caused by	Signs of the disease
Club root	Especially swedes and turnips	Fungus	Roots become swollen and distorted. Plants may be stunted. They may wilt in hot, dry weather and develop purple foliage. Incidence can be reduced by raising the soil pH with lime.
Mildew	Most crops	Fungus	White pustules that produce spores
Rhizomania	Sugar beet	A virus transmitted by a soil fungus	Plants may wilt. Tap roots become stunted. A mass of secondary roots develop, giving a 'bearded' appearance.
Virus Yellows	Beet family	Aphids transmit the virus	Aphids found on the underside of leaves. A complex of viruses cause the leaves to yellow prematurely, which will cause loss in yield.

Main Pests of Root Crops

There are many more pests than the ones I have covered, and you will need an expert in many cases to help you identify your problem. Some species of adult and larvae are similar to one another, so don't panic. Certain varieties of crops will show resistance to the pests, and seed dressed against certain pests gives good control. Insects will probably want spraying, but what product to spray with and how much spray to use must be guided by your expert.

Name of pest	Crop attacked	Caused by	Signs of the pest
Beet cyst nematode	Beet, turnips, mangold, swede, oil seed rape, some leafy brassicas and some weeds	Larvae	Damage caused by small larvae feeding on the roots. Plants become stunted, and outer leaves change to yellow and die.
Beet flea beetle	Beet	Adult beetles	Adult beetles jump like fleas when disturbed. They feed on the surface of the leaves, making small round holes – called 'shot holing,' as it looks as if the plant has been shot with pellets.
Cabbage root fly	Root crops plus cabbage and other brassicas	Larvae	Larvae tunnel into the swollen roots, which ruins the crop
Cutworm	Most crops	Larvae. They are not worms. Cutworms are the larvae of moths, the one that does the most damage is the turnip moth (*Agrotis segetum*)	Brown larvae feed on the roots, causing a great deal of damage. As their name suggests, they cut through the stem at ground level by chewing through it and destroying it.
Leather-jackets	Mainly grassland but sometimes crops are attacked	Larvae	Larvae feed on the stem bases and roots damaging and sometimes killing the plants
Mangold fly	Mangolds and sugar beet	Larvae called leaf miners	The larvae tunnel through the leaves causing brown blisters. Plants become stunted.
Slugs	Many plants	Molluscs	Feed at night making holes in all parts of the plant

Main Diseases of Forage Crops and Oil Seed Rape

Fungal diseases can be sprayed with a fungicide, but this depends on the crop you are spraying and what stage it is at. How much spray and what product you spray with, or if indeed it is worth spraying at all, will be guided by an expert who will probably work for the spray company. Seed can be dressed against diseases such as phoma, and resistant varieties can be chosen for some crops.

Name of disease	Crop attacked	Caused by	Signs of the disease
Club root	Brassicas especially cabbage, kale, cauliflower, swedes and turnips	Fungal infection	Reduced yield. Roots become swollen and distorted. Plants may be stunted and wilt in hot, dry weather and develop purple foliage.
Light leaf spot	Oil seed rape	Fungus	Small white spots can be seen on the leaves, which develop into lesions with pink centres. Stems and pods can also be affected.
Phoma	Oil seed rape	Fungus	Round lesions are found on the leaves. Canker symptoms on stems leading to early ripening and death of the plant.
Sclerotinia – stem rot	Oil seed rape and peas	Fungal infection	Plants wilt and turn yellow. Plants rot at the stem base, and a fluffy type mould, white in colour, may appear.

Main Pests of Forage Crops and Oil Seed Rape

Before spraying insecticide, consult expert help. Seed can be dressed against cabbage stem flea beetle, and some varieties are more resistant to pests than others.

Name of pest	Crop attacked	Caused by	Signs of the pest
Cabbage root fly	Cabbage and other brassicas and some root crops	Larvae	Larvae can be seen on the roots. They eat the roots and destroy them. Plants become stunted, wilt and die.
Cabbage seed weevil and brassica pod midge	Brassicas especially oil seed rape and some weeds	Weevil	Larvae feed in the seed pods, and the other species – pod midge – lays its eggs in the holes. These larvae feed on the inside of the pod wall.

Name of pest	Crop attacked	Caused by	Signs of the pest
Cabbage stem flea beetle	Oil seed rape	Adult and larvae	Leaves are 'shot-holed' by adults and really do look like they have been shot. Larvae also damage stems and leaves. Plants become stunted.
Large white butterfly and small white butterfly	Brassicas	Larvae	Caterpillars feed on the leaves – eating the leaves away and tunnelling into hearts
Mealy cabbage aphid	Oil seed rape, cauliflower, cabbage, Brussels sprouts, swedes and other brassicas	Adult aphids	Whitish or greyish mealy looking aphids feed on shoots and leaves and transmit viruses. Plants become stunted and weak. Leaves turn yellowish white.
Pollen beetle	Oil seed rape	Adults and larvae	Adults and larvae can be seen feeding on buds, flowers and pollen leaving holes in the buds
Slugs	Many species of plant	Molluscs	Eat and make holes in leaves, stems and flowers. Most slugs feed at night. They leave trails of slime.
Thrips (thunder flies or thunder bugs)	Many species of plant	Adult insect	Thrips feed on the leaves causing the leaves to appear silver. May turn flower petals brown while they are feeding. They also leave tiny black dots on the leaves, which are their droppings.

Some pests are quite similar, so you may need an expert to help you identify some of your problems – don't panic.

Main Diseases of Beans

Consult expert help on preventative measures, including spraying and what to do if your crop is attacked.

Name of disease	Caused by	Signs of disease
Chocolate spot	Fungus	Autumn-sown beans are more likely to suffer yield losses. Severe attacks can cause a great deal of damage with the collapse of the stems.
Rust	Fungus	Fungus lives on the leaves. Tiny browny red pustules can be seen on the leaves.

Main Pests of Beans

Consult expert help on preventative measures, including spraying and what to do if your crop is attacked.

Name of pest	Caused by	Sign of pest
Black bean aphid	Adult	The aphids can be seen feeding on the stem and leaves causing damage. The aphids can also transmit viruses.
Pea and bean weevil	Adults and larvae (also attack peas)	The beetles eat small semi-circular pieces out of the leaves, leaving U-shaped notches around the edge
Stem nematode	Nematode	The nematode is microscopic and invisible to the naked eye. They attack leaves, stems and seeds. Plants become twisted and stunted, and later may turn dark brown; seeds become damaged. The pest is seed-borne and can infect and live in the soil for up to 10 years.

Chapter 23

Haymaking and Silage Making

The maids in the meadow
Are mowing the hay
The ducks in the river
Are swimming away.

Hay is grass that is cut and dried in the field, baled and brought into the barn for winter use. Grass stops growing in the winter, and on wet clay land the livestock are better off in the farm buildings to give the land a rest.

I made my first hay as a boy cutting long grass with shears on a friend's overgrown lawn. It was a big garden not quite a quarter of an acre; after cutting, I turned it for 3 days, rowed it up and baled it. To bale it, I found a large cardboard box, cut two slits in the side and lowered strings down the slits to the bottom of the box. I then stuffed the box full of hay and brought the strings to the top, tying them tightly. Finally, I lifted the bale out of the box. It was wonderful hay and kept my rabbits and guinea pigs fed and bedded all winter. Many people have made hay by hand, but I think baling it using a cardboard box is probably unique.

The amount of energy or sugar in grass is highest in the afternoons, and so it is actually best mowed in the afternoon to get the higher energy content. The mower should leave the grass well spread out and not heaped up in tight rows. It is best cut before the grass goes to seed; after this, the quality of its fed value is lower, and dry coarse stems have a low nutritional value.

Fresh grass has a moisture content of about 80%. It needs to be dried to a moisture content of 20%, so that it can be stored in the barn without going mouldy.

Before shutting the field up for hay, it may need some fertiliser but probably not more than 60 kg of nitrogen per hectare with possibly some potash. After cutting, hay will continue to respire and lose nutrients; it needs to be dried quickly, and then fewer nutrients should be lost. The number of times it is turned will depend on the condition of the grass and the weather. If it gets rained on, nutrients will leach out, making the feed quality of the hay poorer.

At baling, some of the crop could be wasted by the baler not picking up all the hay, or the hay might not have been moved far enough away from the hedge and the baler couldn't then pick it up. Most farms with conventional balers use the flat eight system. The baler tows a bale sledge, and when eight bales are collected they are let out of the sledge and left flat on the ground. Later a fore end loader on a tractor or a telescopic forklift picks up the eight bales in one go, and then the eight bales are put on the trailer. You can also unload the trailer using the fore end loader.

Damp hay or hay that is baled too quickly because it is not ready will go mouldy. Hay should smell sweet and be green in colour. However, if very green and above 20% moisture, it may overheat in the stack and cause a fire.

You could get 2–5 tonnes per hectare or more. Conventional bales weigh about 25 kg each, but you can alter the baler to make the bales lighter or heavier, and there would be about 40 bales to the tonne.

Large round bales and large rectangular bales will vary in weight, and some balers make them larger than other balers, but about 500–600 kg would probably be an average weight. The hay is compressed more tightly in these large bales.

Making Hay at Oak Tree Farm

We have made hay on this farm every year since I bought the first field, making over 22,000 conventional bales.

> *If the sun shines on both sides of the hedge on Christmas Day*
> *It will be a good year for corn and a good year for hay.*

We make meadow hay here, not seeds hay made from rye grass. Our meadow hay is a mixture of grass including perennial rye grass, Timothy, cocksfoot, and meadow fescue. We have made some excellent-quality hay over the years, but you need to pick the days with the good weather and no rain. We have never failed, and the crop has never been ruined.

You have to get the weather right, and if you can, you want the grass crop just right as well. Most of the feeding value is in the leaf, and so if you leave it too late, it becomes stemmy and coarse, and will have a low feeding value.

Mowing with older equipment.

Haymaking usually starts here as follows: when we can start will depend on how heavily the field has been grazed, we need enough grass to make the operation worthwhile, but it will be sometime in late May, June or July. The field may or may not have been dressed with nitrogen; if not, it will have had poultry manure spread in the autumn or spring.

Before the crop gets old and stemmy, I will be looking on the internet at weather forecasts for the next 7 days. It usually doesn't take 7 days – perhaps 5 days if the weather is good with no rain. We cut the whole field on the same day – rather than cut in two lots.

Turning with older equipment.

You are lucky to get a reliable weather forecast for 7 days – if I can find 5 or 6 days with no rain forecast whatsoever with plenty of sunshine, I am on the telephone to the contractors, Lindley Hall Farms. They are very good indeed and usually come the day I want them. We then discuss each day and make a decision if we are going to turn it that day, and once we think it is ready to bale they pull out all the stops.

Baling small conventional bales.

In the past, I have asked them to bale some with a big round baler or big square bales, but these are to sell when I have got a bumper crop. I have used these big bales here, but I haven't got the necessary equipment to handle them, which has made things heavy and awkward to handle, so conventional bales are needed here.

When I was a lad, big round bales had not been invented, and in those days British agriculture was ruled by the Ministry of Agriculture, Fisheries and Food not DEFRA. When round bales were invented and were started to be used on British farms, it was rumoured that the 'Ministry' wanted to ban them because they wanted to make sure all the cattle were getting a square meal!

The time it takes to make your hay will depend on:

1. The weather – if it is hot with sunshine all day and a good wind, this is ideal; if it rains, it will set you back.
2. If the crop has young green growth, it will take more making than an older stemmy crop – hay made in May usually takes longer to make than hay made at the end of July.
3. If it is a thick, heavy crop of grass that has had fertiliser or manure put on the field, it may take a lot of making – grass lying thickly on the ground takes longer to dry.

Below I have given two examples of hay making here on this farm:

Top field 4.38 hectares.	Top field 4.38 hectares.
2010 20 June. Mowed. 21 June. Turned. 22 June. Turned. 23 June. Rowed up ready for the baler, baled, carted and stacked in the barn producing 950 conventional bales.	2013 4 July. Mowed. 5 July. Turned. 6 July. Turned. 7 July. Turned. 8 July. Turned. 9 July. Rowed up ready for the baler, baled, carted and stacked in the barn producing 598 conventional bales.
A fairly heavy crop of grass cut in June, but hot windy weather makes it a quick operation taking just 4 days.	Not a very heavy crop – the grass was grazed hard before the field was shut up. Less sunshine and warmth and less wind than in 2010 meant it took longer to make – taking 6 days even though it was a lighter crop.

The number of bales produced over the years has certainly varied from just over 500 to 1800 on this same field.

If you are employing contractors who charge you per hectare for mowing and turning, it is better if you have a heavy crop because a heavy crop will work out at less money per bale. When I worked on farms back in the 1970s, the hay was baled perhaps earlier than today, which was probably more than 20% moisture, and the bales were left out in the field in piles of six or eight. It was thought that if it wasn't quite ready, it could still 'make'

Baling – big bales.

a bit more in the bale, and the rain wouldn't harm it as much in the bale as loose. Nowadays, with better weather forecasts, it is best to bale, cart and stack all on the same day, and don't bale too soon – you don't want a mouldy crop. The old farmers used to say if you can bale some sunshine in with your hay, it will be a better crop.

On a few occasions when carrying in the bales, it has started to look like rain. I remember one year, it was forecast torrential rain with thunder and lightning to arrive during the evening. The contractors arrived at 3 pm to start baling. Half way through, the atmosphere started to change, and we knew the rain and thunder were coming. It was touch and go to get all the crop in before the storm began. The Lindley Hall Farm team always work very hard, and from the start of baling until the last bale is stacked in the barn, it usually takes them only about 6 hours. This particular year, I felt guilty because we had not been out with the teas and coffees – I couldn't let them stop to drink it! We did eventually take them a drink, and they spent only about 3 minutes drinking it, and then it was back to work!

Bringing in the bales.

The last bale was in the barn, and half an hour later the torrential rain, thunder and lightning came. To get it done just in time is a great feeling. One of the best feelings in the world after your successful hay harvest is to lie in bed at night listening to the rain beating on the windows, knowing your barn is full of new hay that is bone dry.

When stacking bales in the barn, I am very particular at getting a straight side to the stack – it has to look as upright and square as a brick wall! I often look over hedges and down farm drives looking at farmers' barns to see how well they have stacked their bales. Years ago, when hay was stacked outside, the bottom rows were placed on edge and not flat. This is because on edge, the sisal bale strings would not be touching the ground and so would not rot. Nowadays, the strings are plastic, and so it doesn't matter which way you stack them.

We stack our bales on pallets in the barn, which keeps them off the floor and stops them getting damp; it also makes it easier to slip some rat poison under the pallets. Hay turns brown when exposed to the atmosphere, only a little way into the bale; if you put some loose straw on top of your stack, the top bales won't turn brown.

Stacking bales in the barn.

Hay bales all safely gathered in.

Silage

Silage is pickled grass. Sugars are converted into lactic acid by bacteria; *Lactobacillus* and *Streptococcus* are strains of bacteria that are anaerobic (anaerobic means without oxygen) and convert sugars into lactic acid, and it is the lactic acid that pickles the grass. The bacteria work really well when the grass is rolled and the oxygen is expelled. I have quite often had surplus grass here, and I have sold the grass to farmers who bring in contractors and make silage out of it.

Mowing for silage.

Wet weather is not such a problem as when you make hay. Silage can be made in cold and cloudy weather; mowing in this area for first cut is usually in the middle of May, but even so you don't want it too wet.

To take four cuts of silage in a year, plenty of fertiliser will be required, and the species of grass will be Italian Rye grass.

The grass is mown, and depending on the weather, the thickness of the crop, the time of the year and which cut it is, it will be rowed up and picked up by the forage harvester probably either the next day or the day after.

The forage harvester picks up the grass, the chopper cuts the grass into a predetermined length, and it is then blown into a trailer.

On small farms with fewer tractors and machines, the trailer to catch the grass can be towed by the forage harvester. On larger farms or farms using contractors, a separate tractor and trailer will run along the side of the forage harvester to catch the grass.

Rowing up.

Silage clamp.

Traditionally, silage is stored in clamps. On the clamp, it is pushed up into a large pile, and the tractor will keep driving on top of the pile to get rid of air. When all the grass is on the clamp, and it is very firm, it can be sheeted with polythene sheeting.

The sheet should be arranged so that it completely covers the grass taking special care with the edges to keep out the air. Tyres will be placed on the top, which help keep it air tight and to stop the sheet blowing off in the wind.

Alternatively, sheets can be purchased to the correct size of your silage pit and anchored down with sand bags, so no tyres need to be used.

The yield for silage can vary from 6 to 18 tonnes per hectare.

Bales of silage have become more and more popular. The silage is baled up usually in large round bales and then wrapped in polythene. It is most important that the bales are not punctured when moving them out of the field or stacking them because if you let the air in, the silage will be spoilt. Vermin such as rats can puncture the polythene; crows can also be a real nuisance, pecking holes in the top of the sheet, and if they are a problem, the bales have to be netted to protect them.

Round bales of silage.

Silage is very palatable to livestock and is excellent for cattle, but I much prefer hay for sheep. Sheep will eat silage, and many farmers feed it to them successfully. I wouldn't feed silage to sheep because if there is soil in the silage or it is mouldy or spoilt, the sheep could get listeria. *Listeria monocytogenes* is a bacterium that causes listeria. The condition is sometimes known as circling disease. Sheep may have inflammation of the brain, go round in circles, and suffer blindness and death. To grow, the bacteria need air and a pH above 5.5 – they may grow if there is a hole in the polythene around the bale or a gap at the edges of the clamp.

How to Make Grass Silage

Italian rye grass is a wonderful, quick-growing, high-yielding grass that is by far the best grass for making silage. It starts to grow early in the spring, and with fertiliser it is unbeatable.

Robert, the best man at our wedding and very-near neighbour, does an excellent job of his silage making and achieves four cuts of silage from each field in the year. He grows and makes the crop as follows. The Italian Rye grass is part of a rotation and follows winter barley; the field is then spread with farmyard manure. The stubble is ploughed in

July and worked down to get a good seed bed, which usually involves power-harrowing twice, and rolled. The soil is analysed to see what and how much plant food is needed. A stale seed bed is then created – the weeds are allowed to grow for about 3 weeks and then a week before sowing they are sprayed with 'Roundup.' After spraying, the field is left for a week and then power-harrowed again. The grass seed is just Italian rye grass with no other grasses or clover at a rate of 37 kg per hectare. Ideally it is sown in the middle of September with no fertiliser in the seed bed. The crop may need some sulphur. Italian Rye grass is a good strong-growing grass and outgrows any weeds that come up. The field is then left until spring, and if it is dry enough in early March it is flat-rolled, which flattens down any stones that would get in the way of the mower and forage harvester.

Ideally, the fertiliser wants to be spread 6 weeks before the first cut on about 1 April depending on the weather; the amount will depend on the analysis results, but it will be mainly nitrogen. Mowing takes place about 15 May, which is completed by contractors and then later that day, the next day or the day after, a forage harvester picks up and chops the grass, which is blown into trailers, taken back to the farm and put into the silage clamp. No additives are added. The field is then given another dressing of fertiliser at about the same rates as the first application, and the second cut would take place around 1 July – traditionally, this used to be Royal Show week, but sadly the Royal Show is now finished.

After the grass has gone, more fertiliser is added – less this time. The third cut takes place about the second week in August, with less fertiliser spread again afterwards. The fourth cut would be baled into large bales; the yield is less, but it is worth doing. This is carried out at the end of September or the first week in October. The fields may be down to grass for up to 4 years.

Haylage

Haylage is cut grass that is turned, baled and wrapped when the moisture content is relatively high. Mild fermentation takes place to preserve the grass. It is dust-free, about 40% moisture and ideal for horses who prefer a wet feed. Once opened, it should be used up as soon as possible in case it goes mouldy.

Grass turned into haylage.

Maize Silage

Maize has become a very popular crop for silage making and is especially beneficial for the dairy herd. It is a crop that is sown late, as maize won't stand frosts and is harvested in September and October. It is an easy crop to grow and benefits from plenty of manure. The seed bed should be very fine, and once the chosen cultivations have been made, and you have a good seed bed, drilling can commence.

A herbicide should be applied either pre-emergence of the crop or early post-emergence to kill broad-leaved weeds and grass weeds. The seed should be dressed to control disease and pests. The crop should be precision-drilled during April and early May at a seed rate of 111,200 seeds per hectare. Approximately 50 kg of nitrogen per hectare should be applied when the crop has started to grow during late May or June. Also, apply a growth stimulant, which is a mixture of trace elements, especially zinc, in early June. The crop is harvested in September and October with a forage harvester.

The clamp should be filled quickly, rolled well to get rid of the air and covered with a complete seal. Maize is a very palatable crop and, as stated, especially beneficial to dairy cattle. Yields can be 60 tonnes of fresh weight per hectare, averaging about 40 tonnes per hectare.

Forage harvester harvesting maize.

Some farmers are now signing contracts to supply maize for anaerobic digesters. In these digesters, microorganisms break down biodegradable material such as maize to produce methane, carbon dioxide and fertiliser. The resulting mixture of gases is called biogas.

Chapter 24

Caring for the Environment

As a boy, I spent my weekends and holidays with my uncle Ben and aunt Doll on a 202-hectare mixed farm. Both my uncle and aunt were real country people, and both had an excellent knowledge of agriculture, horticulture and wildlife. I was encouraged to be a good farmer, a good gardener and a good naturalist.

We watched fox cubs playing and found leverets, stoats, weasels and many species of birds. We carried out a survey of breeding birds on the farm, filling in British Trust for Ornithology nest record cards recording the species, where the nest was found, number of eggs and number of fledglings, finding over a hundred nests on the farm. I was taught wildflower identification and tree identification.

As a teenager, I joined the local bird watching club, attended indoor meetings in the winter and went on coach trips during the spring, summer and autumn to Wales, Norfolk, the Staffordshire moors and many more locations. I had seen and ticked off 175 different British species by the time I went to college aged 19. I took up birding again in 2005 and since then I have travelled over 100,000 miles in pursuit of birds. I have now seen over 450 species of wild birds in Britain.

During college time, I concentrated on my studies – my main course being Rural Studies, and a large part of this course was Environmental Studies. After leaving college, I taught Rural Studies in a secondary school, and part of my course was Environmental Studies, setting up an environmental area in school with British trees, flowers and plenty of shrubs with berries for birds, bird tables, nesting boxes and an ecology pond. I now had no time to continue with birding trips.

Autumn 2013 – the trees have matured on the tree-lined drive.

I planted a tree-lined drive soon after the bungalow was built here at Oak Tree Farm – all oaks, needless to say.

On this farm, I have set up two main environmental areas, the first one a large pond 18.28 metres by 9.14 metres with an island in the centre. I decided I wanted a large pond that would be fairly close to the bungalow so that I could see it easily from the living room, kitchen and master bedroom.

Marking out the pond.

Flattening down the clay to make it firm to hold water.

I marked out where it would go, altering the shape once or twice, and then my contractors Brian and Ben dug out the large hole. It is a clay soil, so once it was dug out, we used a tractor to press the clay down to make a firm bottom and sides that would hold water; I used a spirit level to make sure it was level at the top all the way round – I didn't want the pond to be full at one end and half full at the other.

I couldn't wait for the pond to fill up naturally, and even then it wouldn't have filled to the top, so I filled it with a hose pipe and kept it running; it looked wonderful. I let it settle and was about to stock it with plants and fish when I realised it was obviously leaking. The water was seeping through the clay, so it was back to the drawing board. I then purchased a very large butyl liner from our local aquatic centre. I drained the pond and then lined it with sand so as not to puncture the liner with stones. I put the liner in the very large hole draping it over the island. I then filled it up with water before trimming off the edges, leaving enough edge to bury under the grass, and then cut out the top of the island.

Putting the butyl liner in the pond.

Success – this time the pond held water, and after it had settled I was able to stock it with plants such as Canadian pond weed, water mint, water lilies and irises, and the expensive koi carp.

The pond soon became a magnet for wild pond life such as water boatmen, pond skaters, plenty of dragonflies and birds. Wild mallards nested on the island each year hatching out broods of ducklings. We even had a waterfall.

Finished pond.

I was very proud of the pond, but when our son Jonathan was born, my wife, Sarah, and I decided to fill it in because it was too dangerous with a toddler about and not worth the risk. The pond was filled in, and the area grassed over – you can't tell where it was now, but we hope to dig it out again soon, now that the children are older.

Bank ready for planting trees.

My second large project is still in existence. We have a bank at the end of one of the fields across the road. This bank slopes up to meet the lane, which rises to cross an old railway bridge. The bank was very overgrown with large hawthorn bushes, trees and weeds when I bought the field. Contractors used a digger to dig up everything, and we then made a large bonfire with the bushes.

Contractors and friends, Matthew and David, cut and laid the hedge, and then we were ready to plant the area up with a mixture of British trees. We chose ash, beech, crab apple, maple, oak, rowan, silver birch and some dogwood, which is easily recognised by its bark. They were all planted in 2005.

Hedge now cut and laid, and trees planted.

Bank in winter 2014 showing how the trees have matured.

The trees have since grown well, and now the area is excellent for attracting wildlife, especially birds.

Over the years I have recorded the following 82 species of birds on the farm (some have flown over only):

barn owl
blackbird*
blackcap
black-headed gull
blue tit*
brambling
bullfinch
buzzard
Canada goose
carrion crow*
chaffinch*
chiffchaff
collared dove*
common gull
common partridge*
coot
cormorant
cuckoo
curlew
dunnock*
fieldfare
goldcrest
golden plover
goldfinch*
great spotted woodpecker*
great tit*
green woodpecker*
greenfinch*
grey heron
grey wagtail
herring gull
hobby
house martin
house sparrow
jackdaw*
jay
kestrel
lapwing*
lesser black-backed gull
lesser redpoll
lesser whitethroat*
linnet*
little owl
long-eared owl**
long-tailed tit
magpie*
mallard*
meadow pipit
mistle thrush
moorhen
mute swan
peregrine
pheasant*
pied wagtail*
raven
red-legged partridge*
redstart
redwing
reed bunting
robin*
rook
sand martin
skylark*
snipe
song thrush*
sparrowhawk
starling*
stock dove*
stonechat
swallow
swift
tawny owl
tree sparrow*
wheatear
whitethroat*
willow tit
willow warbler
winchat
wood pigeon*
wren*
yellow wagtail
yellowhammer*

(*Denotes birds that have bred on the farm.)
(**Seen by Sarah only)

My wife Sarah has set up an excellent bird table that is positioned just outside our living room window, and she attends to it daily with various seeds including sunflower hearts, niger seeds and mealworms. The sunflower hearts and niger seeds attract goldfinches, who have brought their fledged young to the bird table. In total there are about 40 birds that visit us on a regular basis.

Mallard duck with ducklings. The duck nested on the island.

The list of mammals seen here at Oak Tree Farm is as follows:

badger (tracks in mud and snow only)
bank vole
brown rat
common pipistrelle bat
common shrew
European hare
field vole
fox
grey squirrel
hedgehog
house mouse
mole
muntjac (tracks in mud and snow only)
rabbit
stoat
weasel
wood mouse

The list of butterflies seen at Oak Tree Farm is as follows:

brimstone
common blue
gatekeeper
large white
meadow brown
orange tip
painted lady
peacock
red admiral
small copper
small tortoiseshell
small white
speckled wood

The butterflies are encouraged with a garden that Sarah has planted that is very butterfly-friendly, the favourite bush being the buddleia, of which we have at least five varieties.

Countryside Stewardship

This is an environmental land management scheme. It is available to all eligible farmers, land managers, owners and tenants in England. It replaces Environmental Stewardship, the English Woodland Grant Scheme and capital grants from the Catchment Sensitive Farming Programme. The scheme is delivered by Natural England, the Forestry

Commission and the Rural Payments Agency. Check their websites for up to date information.

The scheme has three main elements, Higher Tier, Mid Tier and a Lower Tier of capital grants. Generous funding is available and or capital grants for an agreed range of options which are environmentally friendly such as growing winter bird food, providing beetle banks and skylark plots, leaving buffer strips on cultivated land and taking historic and archaeological features out of cultivation and many more options. Therefore the scheme will help the following:

Stubble left over winter will help to bring you an income from the Countryside Stewardship Scheme.

1. Help wildlife and nature including providing food and nesting sites for birds and restoring habitats.
2. Provide pollen and nectar sources.
3. Forestry – by funding and planting new trees and managing woodland.
4. Water / flooding – making water cleaner and reducing the risk of flooding.

Seed firms now provide mixtures of seeds that you can sow around the margins of the fields or in field corners or on buffer strips (a strip of land to protect vulnerable areas such as streams and ponds from fertilisers and sprays). There are many different mixtures of seeds, which may include plants such as field scabious, bird's foot trefoil and many different species of grasses. There are also seed mixtures to create your own wild meadow. Seed mixtures can be purchased for game cover and game-feeding areas – encouraging pheasants and partridges for shooting. Some English fields certainly look different now to during World War II when we 'dug for victory' and as much land went under the plough as possible.

I am not involved with Countryside Stewardship because I only have a small acreage, and I want to farm the land. We have very wide grass verges along the length of the lane that runs parallel to the farm that are good nature reserves, and, as I have mentioned, I have planted up a bank with British trees; we provide food for the birds and nectar-rich flowers for the bees and butterflies; and we will dig out the pond again at a later date.

Woodlands

Woodland Trust

The Woodland Trust is a charity that is not connected with the government. They are happy to help you create some woodland and will subsidise the cost of your trees. Check to see if this financial help is still available.

National Forest Company

The National Forest Company (NFC) is creating the National Forest in Central England covering 200 square miles. If you have a site that falls in the National Forest, you can apply for a range of funds that are very generous, including funding for the creation of new woodland. Check the NFC website for all the up-to-date details.

If you plant trees, it is a long-term project, especially broad-leaved woodland; however, Christmas trees are fairly quick to grow and are ready to harvest after 6–10 years.

Solar Energy

You may be able to afford some solar energy panels, probably starting in a small way and working up to more. Certainly more and more farmers are now setting up solar panels on houses, on farm buildings or in fields to produce some or all of their own electricity, and these provide an income when connected to the Grid. Large solar farms are now in

Solar panels on a farm.

operation including two very large operations in Lincolnshire; one at Bourne consists of 19,760 panels producing enough energy to power 1450 homes per year. Alternatively, large solar energy businesses are looking for land to lease, often for 25 years; they pay for everything and then pay you a rent that is often paid quarterly. Your sheep can still graze on the land after the photovoltaic solar panels are installed in many cases. Rising energy costs have also encouraged farmers to seek planning permission for wind turbines as an additional income when connected to the Grid.

Chapter 25

Progressing to a Mobile Home and Bungalow or House

Mobile Home

I enquired about planning permission, for a bungalow or a house to be built on the farm, to my local county council during 1990 and was told there were various criteria that had to be met before it would be considered. Because I was in a green belt area, it would not be easy. A hobby farm was not enough.

You need to read the government's Planning Policy Statements for Rural Areas online. These change and are updated from time to time, so make sure you have read the latest up-to-date information.

This is what I had to do:

1. show there was a definite need to have someone living on the premises;
2. show that the business provides enough 'man hours' – that is, enough work to keep one person occupied full time per week;
3. prove that my agricultural business was making enough profit to pay myself a full-time agricultural wage or a worker a full-time wage, and support the cost of a dwelling;
4. live in a mobile home for at least 3 years and in doing so improve and enlarge the business.

I went to look at some new mobile homes and agreed to buy one if I got planning permission, so I sent in my application in 1991. I was denied planning permission on the grounds of there being, in their opinion, no real reason for me to live on site; I didn't have enough man hours – only .85 of a full-time person, and so was labelled part time – and I wasn't making enough profit. The planners suggested I should buy a house in the village. They concluded by saying 'There was no agricultural support for the proposed dwelling.' I felt that they discouraged me to try for a mobile home, but I would not give in – I had my heart set on achieving my goal. I decided to work hard for the next 2 years and then try again. I wanted to enlarge the business anyway. I first increased the laying poultry flock from 3000 birds to 4000 birds, and I did this without extending the poultry building. I then bought more breeding ewes, increasing the flock from 30 ewes to 50 ewes.

During this time, I had two break-ins to the poultry egg room: the first time 180 dozen eggs were stolen, and then the second time 720 dozen eggs were stolen, but I did manage to recover some of these, as I found them piled up ready for collection in a neighbour's field. The National Farmers Union were very good and paid out on my insurance.

I was now keeping a large flock of sheep and for 11 years had lambed them indoors in the school farm buildings, but now I had really got too many, and I needed to build a lambing shed and lamb them here on the farm; on welfare grounds, I needed to be able to attend to them not only in the daytime but also at night, but I had to wait until 1993 to have this built.

I was now more experienced with the laying hens, and they were making a good profit, so I made another application in 1992. Secretly I thought it was a good job I had been burgled because I was in fear of being burgled again if I did not live on the premises. I re-applied, and my application was very well prepared to convince the planners. I had objections from people living in the village, but I got accepted to live on site in a mobile home with temporary planning permission for 5 years only. During this time, I needed to work hard at the business and then, getting towards the end of the 5 years, apply for planning permission for a house or a bungalow; or, if the business failed and I made a mess of things, I would have to get out and have the mobile home removed.

I employed contractors towards the end of 1992 to dig out a large rectangular hole, placing hard core in the bottom, making it very firm, and then concrete, which meant I would have a concrete slab to place the mobile home on. Then, a much deeper hole had to be dug for the septic tank. These tanks are made of fibreglass and are concreted in place. Bacteria live in the tanks and break down waste. Outpourings travel by pipe laid in trenches travelling down the field. This had to be agreed with the National Rivers Authority to make sure I wasn't polluting any watercourses.

Foundations for mobile home.

My new mobile home arriving.

My new static mobile home (not caravan) arrived on 2 January 1993 delivered on two lorries, one half on each lorry. I had been to look at them and actually purchased one a few months earlier – I took my parents along, and they were as pleased as I was that I was getting nearer to a house or a bungalow.

The mobile home was manufactured in Wellingborough, Northamptonshire by Tingdene. It was called the Villa and measured 9.753 m × 6.096 m. It consisted of a hall, living room, kitchen, bathroom, one double bedroom and one single bedroom, with an oil-fired central heating boiler. It was fitted out with a lovely kitchen with a refrigerator, a sofa, two arm chairs, a wall storage unit, coffee table, double bed and single bed, bedside cabinets, carpets and fitted curtains – everything I needed except a television and video recorder.

The total cost of the mobile home including transport and siting was £18,570.18; this did not include the concrete base and septic tank. As part of the agreement for me to get planning permission, I had to plant a hawthorn hedge around three sides to hide it. I did plant the hedge, but it seemed to be a ridiculous thing to ask me to do because the hedging plants were about 30 cm tall and would take years to grow tall enough to hide the mobile home.

My mobile home.

I was very happy living in my mobile home – living on the premises certainly made things a lot easier for me. I was still teaching full time, and it was wonderful to return home after a day's teaching to my mobile home with no neighbours making noise nearby – although I realise this doesn't suit everyone.

Bungalow

During the time I lived in the mobile home, I worked hard increasing and improving the business. I purchased 3.57 hectares, which joined the existing field in February 1993, and so I now had a total of 6.73 hectares. I extended the poultry unit so that I could keep 5000 birds. I built a good sheep building/barn and then extended it even further. I then increased the flock to 91 ewes, including some Texel crosses. I was also renting extra grass keep for the sheep so the planners could see I meant business and wasn't just going to build a house or a bungalow and then sell it for a large profit. My heart was in the farm, and I now needed the icing on the cake – a permanent dwelling.

I knew roughly the design I wanted for this dwelling – I wanted a bungalow and not a house, and I wanted the layout to be similar to that of my mobile home. John Bowler Agriculture had a professional architect who worked with them and had a wonderful reputation of securing planning permission on farms with free-range poultry units, and he wasn't too expensive. He advised me and drew up the plans. I know some top architects can be very expensive indeed; architectural technicians may be cheaper, but you need to shop around to see what is best for you. You can design it on your own if you wish, but I would suggest you get a professional to draw up the plans for you. John Bowler also provided me with a letter of support from his company, and the Agricultural Development Advisory Service (ADAS) had written a technical report in my favour. I definitely wanted a brick-built bungalow and not a kit home or a log cabin.

The first lot of drawings were almost good enough, but I thought the windows could be larger. The second attempt was excellent with much larger windows. I chose Baggeridge Moroccan Red bricks and Sandtoft plain small concrete roofing tiles in light grey plus UPVC windows, which would blend in and not stick out like a sore thumb. The bungalow would be constructed next to the mobile home. I would live in the mobile home while the bungalow was being built. The bungalow would have four bedrooms with one used as an office, a living room, hall, kitchen, utility room, bathroom and small separate toilet, with a double garage joining on, totalling 136.27 square metres.

I filled in the relevant forms and applied for Full Application with detailed plans including a location plan, layout plans and details of brick, roof and windows along with my cheque to the Planning Department at the council offices (Outline Planning Permission is just when a farmer or landowner fills in forms and applies to see if it is possible to build on land, e.g. Outline Planning Permission to build a bungalow or a house or perhaps a larger development). The planners are accountable to the local councillors; I can understand they all have a job to do, and I certainly wouldn't like it if a block of flats was built in the middle of a green belt or a monstrosity of a building painted very bright colours.

I realised that if I got planning permission, the bungalow would have an agricultural tie. This means that it can only be occupied by a person who is solely or mainly occupied in agriculture, and so your main earnings or the larger part of your income must come from agriculture. I was teaching full time, my agricultural income was getting close to the teaching salary, and I did explain that I hoped to retire from teaching soon.

The criteria for a bungalow were about the same as for a mobile home – the following formed the basis for my application:

1. I had to prove it was essential for me to live on-site, and to do this I again quoted my two thefts and that on animal welfare grounds I needed to be on-site for lambing. I have never suffered from vandalism, but if I had, I think that would have helped towards getting permission in terms of proof that I needed to live on site.
2. I had almost doubled my man hours with extra laying hens and extra breeding ewes.
3. They knew my business existed, but I had to show it was making a good profit – enough to pay me full time or a worker full time and support the cost of a dwelling. I had to show the planners my books that had come from the accountants, which proved my finances were very sound indeed.

I telephoned all the councillors on the planning committee and even made appointments to see some of them at their homes to put my case forward. I was told all along that the County Land Agent was an advisor to the committee; he would visit me and report back to them, and if he said 'Yes,' it probably would be a yes – he certainly had a great deal of influence. I swept up on the morning of his visit well before his arrival, and from the start we got on very well. He could see that I was a farmer making a real go of things. I was encouraged when he told me that he had visited a farm the day before, and they too wanted to build a bungalow; however, they tried to pull the wool over his eyes by obviously borrowing livestock and also tractors for the day, and he told me they would fail.

The Council for the Protection of Rural England opposed my application, and also there was one objection from a villager that didn't bother me too much. The day of the planning meeting arrived. My parents and I sat up in the gallery at the council offices waiting for my application to be assessed. It was passed very quickly, so quickly in fact that the three of us hadn't really taken it in, and I had to telephone one of the members of the committee later that evening to make sure the plans had passed. He gave me the definite good news that they had passed. The build would be inspected on a regular basis under the Building Regulations.

My plans had been passed for the following reasons:

1. The planning committee stated that the new bungalow would not be detrimental to the character and appearance of the area.
2. They agreed that since the approval of the mobile home, I had shown sustainability of the farm.
3. They stated that I was running the farm at a high standard, and I had improved the quality and appearance of the land – that statement pleased me!
4. The Land Agent had worked out that the farm had a labour requirement of 1.74 full-time persons and passed the financial tests.

There were no objections from: East Midlands Electricity, The Principal Environmental Officer, The District Engineer (i.e. no drainage objections) and The Director of Planning and Transportation (Highways).

The plans were passed subject to certain conditions.

1. Development must start within 5 years.
2. Samples of types and colours of materials were to be taken to the council offices (I had already done that).
3. There must be an agricultural tie in place. A property with an agricultural tie is one in which the dwelling, by law, can only be occupied by persons employed or last employed locally in agriculture and the dependants of such persons, as defined in the Town and Country Planning Act. The Act also defines 'agriculture.'
4. Drainage should comprise two separate systems: surface water from the roof, and foul sewer to a septic tank.
5. I must remove the mobile home within 14 days of moving into the bungalow.
6. The drive must be tarmacked or similar material used for the first 5 metres behind the highway boundary before the dwelling is occupied – this was already done.

Obviously my business had not failed while I had been living in the mobile home, it had flourished, but if the business had failed or was run down I would not have got permission, and I would have to have vacated my mobile home, which would have meant no bungalow.

If you want to build a house on a small acreage, it is no good just growing arable crops such as wheat or oil seed rape because you stand very little chance of getting permission, as there is no real need to be on the premises full time. Also, my business is strictly agricultural – if I had horses or dog and cat kennels, it would not have been passed. On a small farm, you need to be intensive to get enough man hours and enough profitability. It is no good just having three sheep and a donkey.

My plot of land increased in value straight after the planning meeting. Before the meeting, it was agricultural land probably worth about £2000, and now it was a building plot probably worth £60,000. This is another reason why it is difficult to build in the green belt because if you could just go ahead and build on any field, many people would make large profits overnight. Full Application plans last 5 years, so you must make a start on the building work within that time.

If my plans had failed, I could have submitted a new application. The alternative to this is to appeal. The planners have to say why an application has failed – perhaps I could have worked on the failings and amended my plans. I was told that appeals take about 6 months to get an answer and an average of only one in three is successful.

The government is proposing new rules on new dwellings, so before you start you may find that the rules have changed, so you must check with your local planning office.

Getting Organised Ready to Start Building

I got three quotes from local builders for them to do all the work – supply materials and build the bungalow from start to finish. The three quotes were much too high; they certainly wanted to make a good profit out of me. Two said that it would take them 8 months from start to finish, and one said it would take only 6 months; however, it took him 2½ months to send in his quote, which was not a good start and certainly put me off him.

I decided to save money and carry out a self-build, that is myself as project manager; then I would buy all the materials and employ sub-contractors to do all the work.

I have been told by tradesmen on a number of occasions 'Let us buy the materials, Mr Terry, we buy at trade prices and get a better discount than you.' I have found this to be absolute rubbish, don't believe them! I can buy trade and get the same discount as they do, and then I don't add anything on. I purchased most of the materials for the bungalow but not electrical or plaster. On a self-build, you may be able to claim your VAT back on some of the building materials – check with Customs and Excise. I was in the lucky position of being a secondary school teacher at the time. At school, I would say something like, 'David, what does your father do?' 'He's a hairdresser Sir.' 'That's not a lot of good. Adam, what does your father do?' 'He's a bricklayer Sir.' 'Is he really – well Adam, here are a couple of house points for you – ask your father to come and see me, and perhaps I can find him some work.'

Many would call this exploiting my students, but they were 16 years old and in their final year at school. Not only did I find a bricklayer, but this chap had also got a bricklaying partner who, I discovered later, had a daughter at my school who wanted to be a vet (she now is one). I found two carpenters and a managing director of a firm that made roof trusses. He made me an excellent offer: 'Push our Glen at school, and if he gets an A star you will see my prices for the roof trusses. I've got three – a price for the public, a price for the trade and the lowest price ever for you!' Glen got his A star, and I had the most amazing deal.

To buy the materials, I telephoned various firms in the area to see what discount I could get. When given a price, I would ask for more discount or quote a lower figure than I had already been given. I played the hard business man because I had got nothing to lose at this stage – the firm that came up with the best deal was Jewsons.

When employing sub-contractors, you need to know if they are quoting for the whole job (this is what my bricklayer did) or quoting for day rates – I prefer being quoted for the whole job, and you need the quote in writing. Some of my sub-contractors worked almost full time or full time for large main contractors. They agreed to do my work during their holidays, at weekends and after work. You have to be careful here, as the project may take longer than having workers there every day. Most were not purchasing materials for me; they would just provide the labour – I paid none of them up front, and I wouldn't advise anyone to do so – we have all seen the television programmes where good, honest people have paid the builder up front, and he has gone and left the job less than half done.

I found some of the workers from parents of my students at school; one of our farm contractors had a digger so he could dig out the foundations, but I still needed to find ground workers, a roofer, a plasterer, a tiler, an electrician and a decorator.

I know a double-glazed window company owner (I went to school with his sister). I also knew a plasterer well – he was a plasterer who used to farm part time and was now a full-time farmer, and he gave me a very reasonable quote. The plumber I used for the farm lived in the village, so he got the job, and the electrician I chose also looked after the farm electrics. The rest of the tradesmen were recommended by the two bricklayers, so I had now got a full team. I was lucky, and I know it is best to employ builders or sub-contractors that have been personally recommended to you but you might not know anyone that has had work done. I went and looked at my bricklayers' previous two jobs, which were both 10 out of 10, so this gave me confidence in them. It is advisable to see if your sub-contractors are members of a trade organisation, and check to see if they guarantee any of the work – they probably won't, but the electrician and the plumber may well do.

I realised that some things would probably go wrong, and although prices were agreed, I did speak to the 'subbies' and discussed with them about what would happen if I had a real major problem that was totally unexpected or in fact I decided on some extra work, and how I would go about paying for it. They said they would be fair and cross that bridge when they came to it. I did have some extra jobs that I got my bricklayers to do: a brick fireplace in the living room, two brick pillars at the front entrance, and some brickwork for a fountain in the front garden. We negotiated a very fair price for these jobs. Before you start work, check that your Employer's Liability insurance and Public Liability insurance are up to date. Your new house can also be insured for structural damage – check with your insurance company to see what is available.

Starting Work

I was looking forward to being the project manager, which would mean I would be the chief organiser – planning ahead to get workers here in sequence, starting with digging out the foundations and then getting the next crew in as quickly as possible, rather than having the building site standing still with no activity. I was teaching full time, but I could talk to and check on the 'subbies' before and after school. They started work at 8 am; I started work at 8.45am. I could come back in my lunch hour to check, and I could come as soon as possible after school. On more than one occasion, I was rushing about in my lunch hour purchasing extra drainage pipes and timber – returning with them before the afternoon session of school began and not having any time for lunch. I kept a diary, which is advisable for your own records, but also if there is any argument with your 'subbies,' they can see that you have written things down. I also kept the plot tidy.

Foundations

Work started on 29 March 1997, and my aim was to be moved in by Christmas 1997. It was a brilliant feeling marking it all out with my tape measure on a plain grass field. I had my photograph taken digging out the first spadeful of soil – or the first 'sod.'

The first spadeful of soil 29 March 1997.

I was lucky to have a level green field site with easy access for a digger, a good drive for the material to be delivered and plenty of storage space, which is important for your materials – it is no good stacking your roofing tiles in front of your bricks.

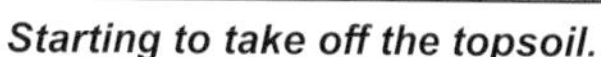

Starting to take off the topsoil.

Digging out the foundations.

Brian came and scraped off the topsoil with his digger, and then the two of us used lime to mark out the exact position of the walls; then he dug out the trenches for the footings.

The trenches were dug out, which took only a day. I paid Brian by the hour – be careful here because some digger drivers want paying for travelling to your site; check this, and if they live a long way from your site, forget them!

I telephoned and organised my next crew to arrive – they were the excavation firm. I ordered the concrete after getting three different quotes for the same mix and got delivery times sorted (which they don't always adhere to). Twenty-eight cubic metres was poured into the trenches.

With the excavations team was a much older labourer aged about 70 and a young man aged about 18. The younger person made fun of the 'old boy' and kept on saying he wasn't strong enough to do the work anymore. The wise 'old boy' said 'I can wheel something in a wheelbarrow over to that poultry shed and you won't be able to wheel it back I bet you £20.' 'You're on,' said the youth. The 'old boy' fetched the barrow and said 'Get in son!'

I ordered the bricks and the breeze blocks – the council had agreed the colour of the bricks at the planning stage. I ordered more bricks than I needed in case we ran out. There was a house built in the village just before mine, and three-quarters of the way through the build they ran out of bricks. The new bricks that were purchased were a slightly different colour to the ones that were first laid, and to me it stands out a mile. I also needed more bricks because I intended to be posh and have two brick pillars built at the drive entrance, and I also wanted a brick fireplace.

Starting to Lay Bricks

The first bricks being laid.

When a bricklayer goes to a house to give a quote for a new extension, what are his first words to the lady owner? 'Tea with two sugars, love!' I hired the cement mixer so that work could begin. Colin and Tom, and their labourer, 'Destroyer,' really are the most professional team anyone could wish for. Colin was not only a bricklayer but also my advisor, and I could telephone him at any time to discuss any building question. The first day of the bricklaying was 18 April 1997; the task was to start to lay just three rows of concrete blocks and five courses of bricks. They all made a good start: two men working 10 hours with one labourer laid 1000 concrete blocks and 800 bricks in the day in total.

Drains and Floors

The ground worker subbies told me what to order and how much to order. Drains were laid using plastic drainage pipes with man holes incorporated to make it easy to unblock drains if need be at a later date. Foulwater is water from the toilets, bath, shower, washing machine and dishwasher. All drainage pipes were laid on a bed of pea gravel. To catch the rainwater, you obviously need guttering and drainpipes, but we were not ready for these yet. We dug the soakaway for the drainage pipes to run into. The soakaway is simply a large hole filled with hard core about 5 metres from the house.

Ready-mix concrete delivered to make the floors.

All drains fall away with the least number of bends in as possible. Once the drains were laid, we could then concentrate on the bungalow floor. First, it was compacted. Then a layer of sand was added and also compacted. Then plastic sheeting and polystyrene were added for insulation, and metal reinforcing grids were added to strengthen the concrete base. Finally the concrete was added.

Nowadays, things are completed differently, and the concrete has fibreglass fibres in it to make it stronger, which cuts out the need to use metal reinforcing. It is tamped by hand, and then a machine takes a lot of the hard work out of the job, troweling it up for you.

All the time the drains and floors were being done, I got the carpenters to construct dummy doors and windows. These were made to the exact size of the doors and windows, and were put in place so that the bricks could be built up to them. If you put your best windows in at the start, they may get damaged.

Bricklaying

You have to construct an inner skin, which is block work, plus an outer skin, using the best bricks, with insulation material in the cavity. You can use timber stud work (this is usually completed upstairs) on the inside dividing room walls, but I preferred to use the stronger breeze blocks. A standard block is equal in size to six standard bricks and measures 440 mm × 215 mm, and so it can be built up very quickly.

I personally think that you can't beat looking at a brick-built house or bungalow being constructed. The more I saw the bricklayers working, the more I realised they were two of the best in the business. You only have to look at the photographs of their work to see how spot on they are. Their labourer, 'Destroyer,' was genuinely nicknamed thus because he was always asking for a 'sub.'

The price of bricks varies greatly – the hand-crafted ones are the most expensive. I had chosen a very nice brick – not hand-crafted but not a cheap one either. You can choose a cheap brick and then render, but this was not for me. I didn't want my lovely bricks covered up. If you lived in an area like the Cotswolds, your house would probably have to be built of local stone, which is more expensive than brick. The planners probably wouldn't let you build a brick house in this sort of area.

The bricklayers soon needed scaffolding to get up to roof height, which saved a lot of reaching and stretching; again, I got quotes from different scaffolding firms and chose the cheapest.

Roof Trusses

Glen's father had been and measured up for the roof trusses – they were made at his factory and were delivered already made up. The two carpenters came and fitted them in position in the rain, fixing them very quickly, and they all fitted like a glove with none left over. The disadvantage of my roof trusses is that they don't give you enough space to construct a room in the loft very easily, you would just manage one long room that would look like a corridor.

Brick laying completed and roof trusses fixed.

Guttering was fixed but not the down pipes yet. Once the roof trusses were fixed, I was ready for the tiler.

Tiling

One of the best (if not the very best) roofing tilers in this area is Gary from Atherstone, Warwickshire. He is the absolute master. I was told before I started this project to get first-class bricklayers and tilers because it is their work that you mainly see. Gary agreed to do the job, so I was lucky to have both sets of good men. I purchased all the roofing materials, so Gary just had to provide the labour. He arrived with his helper to prepare the roof for the tiles. Roofing felt goes on first and then the battens, which the tiles are fixed to. I was amazed at the way Gary walked up the ladder as straight as a gun barrel with tiles on his shoulder and never a grumble or a moan.

Tiling the roof.

You can get handmade clay tiles, but when anything is 'handmade' it is usually expensive.

Windows

The double-glazed window firm had made up all the windows, and the dummy windows had been taken out well in advance. The UPVC double-glazed sealed units were my choice; they are maintenance-free with no painting required. I went for the white windows with a UPVC back door but a solid mahogany front door. I could have purchased soft-wood timber windows or the most expensive hard-wood timber windows, but I didn't fancy the maintenance. I am not a fan of Georgian windows, which have extra glazing bars, and I think the small pane effect spoils your view looking out.

Windows fitted at the rear of the bungalow.

Carpenter

Now that the roof was on and the windows and back door fitted (no front door yet), we were ready for the first-fix carpentry, which included fixing the door frames, but not the internal doors (this was completed by Pete and his helper), and finally the front door.

Electrician

The first-fix electrical work came next. I won't touch electrics; I know nothing about the subject, so I just left them to it. They supplied all the materials and installed what seemed like hundreds of metres of cables that were left fixed to the wall and terminated at the metal boxes ready for the plasterer.

Plumber

The first-fix plumbing was completed by Paul, who is an excellent chap and countryman, and has come out on numerous occasions at very short notice when I have had a plumbing problem on the farm – often coming out as quickly as a vet would. Paul installed all the pipe work, and we were then ready for Kevin, the plasterer.

Plasterer

Kevin the plasterer fitted the plaster-board ceilings and then put a skimming of plaster on top of them and plastered all the walls with a first base coat. He then came back and gave the walls a second coat. I was amazed how Kevin could mix up his plaster on a table in the centre of the room and then run very fast with the plaster to get it to the walls.

Plaster needs to dry out before decorating takes place, and when you do decorate you might get a few cracks appearing afterwards; this is nothing to worry about, and the cracks can be filled in when you decorate next time. Nowadays, most houses have dry plaster boards fixed to the walls and then a skimming of plaster on the top.

Second-Fix Carpentry

Once all the plastering was completed, we were ready for the second-fix carpenter, electrician and plumber. The carpenter's work included fitting the front door, internal doors and skirting boards. They also assembled and fitted the kitchen and the bedroom wardrobes.

To make the bungalow look 'rural,' they fitted beams in the living room, kitchen and utility room. Before they were fixed to the ceiling, they were 'attacked' with an axe to make them look old, and then they were stained dark brown.

Second-Fix Electrics

Again, I left the electricians to it. Work included fitting the plug sockets, light switches and wall lights.

Second-Fix Plumbing

Paul was soon back, and work included fitting the heating system – boiler, hot water tank and radiators – and then testing the system. He fitted the bathroom out with bath, shower, toilet and bidet plus a separate toilet and washbasin in the small toilet off the office. Work in the kitchen and utility room included fitting the sinks, washing machine and dishwasher.

The bricklayers came back and built my brick fireplace and two brick pillars at the entrance to the farm drive, completing them with two large stone balls on the top. The work continued with electric garage doors fitted and down pipes fitted to the guttering, and then it was the turn of the decorator.

Decorator and Tiler

The decorator came in, and I instructed him to do a very straightforward job – white skirting boards and window ledges, and magnolia emulsion on all the walls. My father stained all the internal mahogany doors.

My Dad painting wood stain onto the mahogany doors.

I purchased all the kitchen tiles – both floor and wall, and enough tiles to tile the whole bathroom, and contractors came and completed the whole job.

Carpets, Curtains and Furniture

My Mum fitting the curtains.

I chose 'rose' Kosset carpets to be fitted throughout the whole bungalow and got a tremendous discount on this. My mother fixed all the new curtains in position and made some cushions to match.

I purchased all new furniture, with my parents buying me a kitchen table and chairs, and dining room table and chairs. The building inspector passed it fit to live in, so I was ready to move in. Moving in was easy because the mobile home was just next door. I moved in during mid-November 1997; it had taken 8 months to complete, which is about the same time professional builders would have taken. The total cost was £54,000, including the fitted kitchen and fully tiled bathroom. I had arranged to sell the mobile home, and this was collected 2 days after I had moved in – I sold it for £11,000.

I move in.

Finished bungalow.

The drive was improved and fenced, and it was lined with oak trees. However, it didn't end there: a front garden was created with lawn and roses, and a rear garden was created with a lawn, heathers and dwarf conifers. A water feature built with pond and fountain, and the porch filled in with a door and glass. In September 2000, the drive was tarmacked.

View from the top field.

I really enjoyed being the project manager, and at the time I thought I could do another one of these for somebody else, but I never had the opportunity. Don't let anybody else do it – you save money by doing it yourself, and it is very enjoyable.

Tarmacking the drive.

Oak Tree Farm in spring.

Chapter 26

Diary of Events and Achievements

When, at the age of 18, I went for my interview on a farm, at the end of the interview the farmer said 'If you break anything or you forget to do something, you must tell me – we have no secrets here because the potatoes have eyes, and the corn has ears.'

1989

26 June: 3.164 hectare field purchased at auction costing £21,000. It hadn't been farmed for a number of years, and it was smothered in weeds.
August: Field sprayed with 'Roundup' and hedges cut.
September: Field cultivated, grass seed sown, field fenced and gated.
November: I was elected President of the Kerry Hill Flock Book Society.
December: I became a Chartered Biologist and member of the Institute of Biology.

1990

May: Sheep-handling pens constructed.
June: My first crop of hay harvested.
July/August: Free-range poultry house built for 3000 laying hens complete with egg room, bulk feed bin and chain feeder costing £30,000. Three thousand point-of-lay pullets delivered. My Mum and Dad collect the eggs during the afternoons while I am teaching for over a year – with no wages! They were lovely parents, and I certainly wouldn't be where I am today without their love, help and support.
All summer: Grass grazed by my 25 Kerry Hill ewes and their lambs (wintered at school).
August: Pullets start to lay.
September: East Midlands Electricity bring electricity on to the farm – total cost including connection to the mains: £6000.

Mum and Dad sitting on the steps of my mobile home.

1991

July: Homebred Kerry Hill ram lamb 'Dale' wins first prize at the Royal Show and first prize at the Royal Welsh Show.
Second crop of hay harvested, and hay has since been harvested at this farm every year.
November: I was awarded the cup and glass tankard for achieving the Champion Flock of Kerry Hill sheep in Britain – the highest award you can win with Kerry Hill sheep.

Champion Flock of Kerry Hill sheep in Great Britain.

1992

July: Awarded First Prize and the Best Group of three Kerry Hill sheep (one ram and two ewes) winning the John Terry cup at the Royal Show.
October: Laying hen flock increased to 4000 birds.
Flock of Kerry Hill breeding ewes increased to 50.
Plans passed for a mobile home.
November: Water pipes brought from the village and mains water connected costing £6000.
Concrete base made for mobile home, septic tank installed.

1993

2 January: My new mobile home arrives costing £18,570.18, and I move in.

February: 3.57 hectare field purchased privately costing £23,000 (the whole field is actually 4.38 hectares, but the other 0.81 hectares furthest away were not for sale). This new field joins on to the existing field and is to be called the 'Top Field.'

April: I was elected by the Council of Awards for Royal Agricultural Societies as Fellow of the Royal Agricultural Society (FRAgS) in recognition of outstanding service to and achievement within the agricultural industry. This is like an agricultural OBE - it was for my work teaching Agriculture at school, and I was very proud to accept it.

October: A new sheep building/barn constructed costing £5000 for materials and labour.

1994

April: Another new building added to the sheep building/barn costing another £1000 for materials and labour. Not all the lambs have mothers, but most do. A few are bottle-fed.

My Mum bottle-feeding a lamb.

September: I purchased a Texel ram and 25 Texel cross ewes to breed commercial sheep to sell as finished lambs.

October/November: The free-range poultry house extended by approximately another third costing £5000 for materials and labour.

1995

January: The laying hen flock increased to 5000 birds.

March: The first lot of Texel cross lambs born.

August: Texel cross lambs sold at market.

1996

3 May: 3.144 hectares purchased privately costing £25,000. This field is situated directly over the road from the existing two fields.

September: The new field across the road sprayed with 'Roundup,' ploughed, power-harrowed and grass seed sown.

Plans have been passed for a bungalow.

1997

March: The new field across the road fenced and gated.
29 March: the first spade of soil dug out for the new bungalow.
Spring/summer/autumn: new bungalow built.
August: Brick pillars built and new gate fixed to the farm entrance.
November: The new bungalow is completed, and I move in. The total cost is £54,000, including fitted kitchen and fully tiled bathroom.
Bungalow completely furnished.
Mobile home sold for £11,000.
Front garden created, drive in front of the bungalow measured, dug out with hardcore and road planings added. Oak trees planted on either side of the drive.
December: Post and rail fence erected on either side of the farm drive.

1998

May: New rear garden created at the bungalow.
July: The floor is concreted in the poultry house – the floor up until now was soil.
August: Front garden improved and water feature added.
September: A large pond dug out at the rear of the bungalow.
Pond planted up with water plants and stocked with koi carp.
October: I purchased my first Derbyshire Gritstone sheep – three ewe lambs and a ram. (I am still keeping, breeding and showing the Kerry Hill sheep.)

1999

May: The large pond was not holding water, so the pond was drained and a liner added costing £800.
July: My first attempt at showing Derbyshire Gritstone sheep at the Royal Show achieving 1st prize and Reserve Breed Champion.
December: Retired from teaching after 25 years' service.

2000

6 March: I purchased the 0.80937 hectare field, which joined my Top Field. This land was originally part of the Top Field and was not sold when I bought it. However, I purchased it at last at auction for £22,000, which was expensive.
May: Mallards breed and hatch 8 ducklings – nesting on the island in the middle of the large pond.
August: The whole of the drive tarmacked, with kerb stones concreted in place.

Achieved Supreme Interbreed Champion sheep at Canwell Show near Birmingham, beating every sheep in the show with a Kerry Hill ram.
My 20th year showing sheep. A Kerry Hill ewe, Oak Tree Belinda, was Reserve Champion for Any British Breed at Hope Show, Derbyshire, winning my 1000th rosette.

2001

February: On 20 February 2001, Britain's first outbreak of foot-and-mouth disease for 20 years was confirmed. This farm escaped the disease, but it was confirmed on three farms very close by.
October: The sheep building/barn was fitted out with new metal hayracks and new gates, and the floors were concreted.

2002

July: Awarded first prize and the Best Group of three Kerry Hill sheep (one ram and two ewes) winning the John Terry cup at the Royal Show for the second time.
September: I have sold all the Texel cross ewes and the Texel ram so that I can concentrate on breeding the Kerry Hills and Derbyshire Gritstones.
October: I judged the Kerry Hill sheep at the National Kerry Hill Sheep Show near Rotterdam, Holland.

2003

July: I judged the Kerry Hill sheep at the Royal Show, Stoneleigh. This is the highest accolade to judge here – perhaps the equivalent of a football referee refereeing the Cup Final. This has been an ambition of mine for a long time.
September: I exported 15 homebred pedigree Kerry Hill sheep to Holland (six ewe lambs, seven ewes and two ram lambs); they went to new homes on three farms.
I became the happiest man in Great Britain – meeting Sarah and knowing I will spend the rest of my life with her. Sarah, a teacher and very much interested in the countryside, agriculture and horticulture, means I have at last found my perfect match. I am very lucky – Sarah becomes involved with the farm.

Judging Kerry Hill sheep at the Royal Show.

Two rams – their semen exported to California.

September: Kerry Hill semen from two of my Kerry Hill rams exported to California. The semen to be used on Cheviot ewes to 'make' Kerry Hill sheep. Cheviots are very similar to Kerry Hills with upright ears but without the black and white markings. The first generation of lambs will be 50% Kerry Hill and 50% Cheviot. Then more semen will be used to get the next generation, and then this lot of lambs will be 75% Kerry Hill and only 25% Cheviot, and then breeding will continue until finally we get pure-bred Kerry Hill. Some semen from the same two rams is now stored in the Rare Breeds Survival semen bank.

2004

July: Achieved another ambition – the Supreme Champion Derbyshire Gritstone sheep at the Royal Show with 'Charlie.'

Charlie.

The Atherstone Hounds.

August: New plastic floors and new automatic nesting boxes fitted in the laying house.
December: The Atherstone Hounds meet for the first time here.

2005

16 January: Jonathan born, 9 lb 7½ oz. Our first child. Sarah and Jonathan both fit and well.
October: The large roadside bank cleared, the hedge cut and laid, and trees planted.

2006

July: Awarded Supreme Champion Derbyshire Gritstone sheep at the Royal Show with 'John Wayne.'
Kerry Hill semen from Oak Tree Charles and Oak Tree Teddy exported to Virginia USA to be used on Cheviot ewes to 'make' Kerry Hill sheep eventually.
August: Awarded the Champion of Champions prize with a Derbyshire Gritstone ram (Charlie) at Hope Show, Derbyshire. This is equivalent to British Champion. This competition is open to the Supreme Champions male or female from all the shows around the county holding breed classes.
Filled the large pond in: ponds are dangerous places for children.
September: Constructed a 'lean-to' on the side of the poultry house. This is a scratch area for the hens, which is now a legal requirement. It cost £1800 for materials and labour.

Champion of Champions.

2007

February: Sarah wins Champion Light Sussex in Show with one of her cockerels at Stratford upon Avon Poultry Show.

July: Supreme Champion Derbyshire Gritstone Sheep at the Royal Show with a ewe – 'Janet.'

18 December: Roseanna born, 9 lb 1 oz. Our second child. Sarah and Roseanna both fit and well.

2008

July: Supreme Champion Derbyshire Gritstone Sheep at Royal Show with a ram – 'John Wayne' (so called because he is enormous, confident and full of himself, and swaggers when he walks).

November: I retired from being a council member of the Kerry Hill Flock Book Society after serving for 21 years. I was presented with a wonderful shepherd's crook, which I treasure.

2009

February: Acquisition of 3.273 hectare field, which we purchased privately costing £77,500. This new field joins onto the other field across the road. This new field is the 'Arable Field.'

Hedges cut back in new arable field and ditch cleared out.

July: Supreme Champion Derbyshire Gritstone Sheep at the Royal Show winning this award for the fifth time (John Wayne's third time). He also won the Supreme Champion Exhibit Any other Breed MV non-accredited. In the Supreme Interbreed Championships, he was one of the eight finalists out of 921 sheep – a Texel ram won. This was the last Royal Show,

John Wayne at the Royal Show.

and we couldn't have wished for a better result.

July: I judged the Kerry Hill sheep at the Royal Welsh Show at Builth Wells.

August: Jonathan shows his first lamb at Poynton Show, Cheshire, achieving 2nd prize aged 4 with encouragement from Sarah and myself.

September: Arable field cultivated, power-harrowed and winter wheat sown.

2010

August: The first wheat harvest, achieving 8.98 tonnes per hectare and sold for animal feed.

October: Wheat drilled for the second time on the arable field.

2011

April: A complete new sheep-handling system installed costing £2779.

August: The second crop of wheat harvested.

Winning the Interbreed Sheep Champion at Poynton Show, Cheshire, with a Derbyshire Gritstone ewe – Oak Tree Emma – and Jonathan achieves first prize with his lamb aged 6.

New handling system.

September: The arable field cultivated, power-harrowed and sown with oats.

November: One of Jonathan's hens 'Snowy' lays 102 eggs in 102 days! He has a report in a national poultry magazine.

2012

March: Jonathan lambs his first ewe. Sarah and I were very proud, but it did mean his clothes had to go straight into the washing machine and him into the bath.

July: New nipple water drinkers fitted in the free-range laying house replacing the old bell drinkers – materials and labour cost £2249.36.

I judge the Kerry Hill sheep at Castlewellan Agricultural Show, County Down, Northern Ireland, travelling from Birmingham airport to Belfast airport. This combines business with pleasure – involving the whole family away for the weekend.

August: My first crop of oats harvested, which yielded 7.75 tonnes per hectare – well above the national average. Jonathan, aged 7, wins the Supreme Champion Derbyshire Gritstone award at Poynton Show with his ram lamb 'Jonathan' - beating me. He then appeared in various newspapers as well as on Central Television.

1 September: I was awarded the degree of Bachelor of Arts honoris causa by the University of Worcester.

September: My first appointment to judge Derbyshire Gritstone sheep at Hayfield, Derbyshire.

Jonathan with his champion ram lamb.

2013

March: Jonathan and Roseanna now exhibiting at poultry shows – Jonathan showing his large fowl Brown Leghorns and Roseanna showing ducks and eggs – both of them winning Championships. Roseanna is helped by Sarah, who is brilliant at picking out the best eggs and is teaching Roseanna how to do it and what to look for.

June: Jonathan is breeding his own Derbyshire Gritstone sheep and Brown Leghorn large fowl, and Roseanna is breeding Call Ducks.

A new roof is put on the free-range laying house, costing £22,000.

October: Jonathan judged the class for under-25-year-olds showing ram lambs at the National Derbyshire Gritstone Show and Sale at Clitheroe, making him (aged 8) probably the youngest sheep judge in Great Britain. A newspaper article was written about him. Roseanna has shown eggs at three shows this year, achieving Champion Eggs in the Show at all three shows. Sarah and I are proud to see our children involved with the farm.

November: I have been selected to judge the ewes at the National Derbyshire Gritstone Show and Sale in 2014 and the rams in 2015.

2014

February: Sarah buys some Serama bantams - the smallest breed of chicken in the world - with the intention of breeding and showing them.

April: Spring barley sown on the arable field (sown late in the year due to wet weather).

May: Sarah's first Seramas are bred.

June: News from the USA. As expected, Kerry Hill semen that I

Sarah and Roseanna with the Serama bantams.

exported a number of years ago has been inseminated into Cheviot ewes for a number of generations, now resulting in almost pure Kerry Hill sheep being bred. Next year we will hopefully see the first ever pure Kerry Hill sheep in the USA.

July: Our son Jonathan purchases four pedigree show quality Dorset Horn ewe lambs from Wiltshire to start up his own flock, which he will breed and show.

Jonathan's Dorset Horn sheep.

August: Jonathan's home-bred ram achieves Champion Derbyshire Gritstone sheep at Bakewell show, Derbyshire, with me achieving the Reserve Champion with a ewe. I don't mind being beaten by my own son! He also entered 3 shows with his Dorset Horns and won first prize with them at each show. We now look forward to him breeding his own. Both children have attended poultry shows – Roseanna achieving championships with her Call Ducks and chicken eggs, and Jonathan championships with his Brown Leghorn large fowl.

September: First crop of spring barley harvested.

October: I judge the ewes at the National Derbyshire Gritstone Show and Sale at Clitheroe, Lancashire.

Sarah shows the Serama bantams and wins first prize.

Up to Date

Kerry Hill sheep have been kept and bred at this farm for the whole time and Derbyshire Gritstones since 1999 selling stock all over Great Britain including a Kerry to Kerry in Powys and a Kerry to the Shetland Islands. Both breeds have won a total of over 1600 rosettes, which are fixed to our wall in the utility room.

Laying hens have been kept since 1990, laying probably about 30 million eggs over the years. If laid end to end, these would reach over 1100 miles, which is more than from John O'Groats to Land's End.

We have farmed the land to the best of our ability, and at the moment there are 25 Kerry Hill ewes, 35 Derbyshire Gritstone ewes, four Kerry Hill rams, six Derbyshire Gritstone rams and 5000 laying hens, and we have one field of wheat.

The land totals 13.92 hectares – 3.273 hectares is arable; the rest is grassland and farm buildings.

Remember – It's nice to be important, but it's important to be nice!

Chapter 27

Conclusions

Ariel view of Oak Tree Farm showing the bungalow, farm buildings and some of the land.

Hopefully, you will have read my book, or at least some of it - so what will you do next? You might say it's alright for John Terry because he started his project years ago, and it was easier then. It probably was, but I didn't think it was easy at the time, and I am sure not as easy as before my time. Land has got more expensive in comparison with other things, and that today is the difficulty – finding enough money to buy. Family farms have maintained the landscape as well as produced food for generations and helped keep rural communities up and running along with the village school, village shop and pub. I like land to be farmed well and the produce made the best of. I find it depressing that in the UK, seven million tonnes of food is thrown away each year – I hate to see waste. However, my advice is to get up and have a go; - 'He who dares wins!' Get up off your backside and achieve something that you will enjoy. It won't be easy, and you will have to work long, unsociable hours, missing out on holidays and days off. You really need to be physically capable; large farm animals are strong and can knock you over. You need to be able to lift bales of hay and straw, and bags of feed, and move barrow-loads of muck. You will be working on your own for a great deal of the time with no company and so no one to talk to – working in rain, hail, thunder and lightning, snow and hot sun, and working 365 days a year with few holidays. I know many people are certainly trapped in factory jobs, and some will have a boring work life and long for the countryside but can't get out of the 'rat race' owing to a heavy mortgage and loans. Some, however, will come into agriculture as a second career and have a job involved with agriculture such as a sales representative, and just a few may start farming on their own. On the other hand, some established farmers can't make their farm pay and so need to go out to work part time or full time or perhaps carry out a bed-and-breakfast business on site or let buildings for light industry or barn conversions for dwellings.

I am sure I did things the right way round. I left school at 18, worked on a farm for a year and went to teaching college for 3 years. My school farm progressed, and I kept my own flock of sheep at school. I have given at least 500 talks and after-dinner speeches, I have written four books and lived with my parents, and was too 'tight' to buy a Mars bar with my coffee at school break time and so saved my money. Once I had built up a farm, I could then afford a wife and family. If I had got married at 18 and fathered numerous children, it would have been a struggle, and I probably wouldn't be where I am today – but who knows, I could be wrong.

I know of a couple living locally; he has worked on farms all of his life, and they have attended probably a hundred sales of small parcels of land over the last 25 years. They were at the auction when I bought my first field; each time they have been hoping to buy, but they have never taken the plunge. Either it has been too expensive or the land was too wet, or there was no road frontage, or it wanted too much money spending on it, or they didn't like the neighbours. They are now getting too old, but I believe they are still looking for land. It is no good when you are about 80 saying 'If only I had rented that field' or 'If only I had bought that paddock' or 'I wish I could have kept some pigs or chickens.'

I often go into a local shop, and one of the shop assistants who was in my class at school will often say 'I should have married a farmer.' She is a 'townie' living in the middle of a city and is longing for the country life. The nearest she gets to the countryside is watching Emmerdale on the television. However, I do realise that British agriculture has always faced challenging times with often low prices for your arable crops and livestock, increasing feed and energy costs, outbreaks of disease such as foot-and-mouth and DEFRA's endless rules and regulations and forms to fill in.

Check the websites of colleges that offer courses in Agriculture – depending on your circumstances, you may be looking at full-time or part-time courses, but I would encourage you to do at least one course. Many of my former students have completed courses – some are involved in farming, and some are not. Many went to work on farms after college but have given up the farming life to work in industry to earn more money, especially when they get married and have children. Many of you will have worked on farms and have gained practical experience. You can be a hobby farmer, a part-time farmer or a full-time farmer.

Hobby Farming

You still keep your main job, and to be a hobby farmer is easy – if you have a garden, it may be possible to keep a few laying hens or a few ducks or, if you clean out your garden shed, a goat – assuming you have tolerant neighbours and deeds to your house that do not state 'no livestock.' You don't have to start at a certain age; you can start when you are 6 or 7 years old with your parents and then sell your eggs to your parents. I did that aged about 14. If you have really tolerant neighbours, you might even be able to keep a cockerel with your hens, either large fowl or bantams; you can then breed them, show them, visit shows at weekends and mix with like-minded people, and have a thoroughly good time.

Winter at Oak Tree Farm

Hobby cattle, sheep, pigs, geese and turkeys are all possible, but you need some more extensive facilities than a back garden. I know you can construct a shelter from straw bales and tin sheets, but in a gale they might not stay put, and electric lights are a wonderful thing, certainly beating someone holding a torch outdoors on a stormy night while you are trying to lamb a ewe or even worse holding the torch as well as lambing the ewe.

The secretary of our local poultry club rents eight allotments that all join up and cover about an acre. He grows vegetables and keeps fowls, turkeys and geese. He has constructed pens and runs, and it is a successful operation. However, not all allotments will let you keep livestock, and you need to be careful of vandalism and theft when livestock are kept.

Part-Time Farming

This is farming to fit in with your life and job. To be a part-time farmer and hopefully make some money, you will be thinking of farming on a larger scale than hobby farming. I really do think the very best way to start is to take some grass keep buying store lambs in August and September, rearing them through the winter, and then selling them as lambs for freezers – you need no buildings for this. To take grass keep is the bottom rung of the farming ladder, and you need very little capital. Dairy farmers are often keen to see their grass grazed off, and it is an extra source of income for them. You will see grass advertised in newsletters from livestock auctions and on the internet, and also advertisements in agricultural magazines. It is possibly easier to rent a pony paddock and stables than rent a field for agriculture. Land for ponies can command a high rental premium – probably too high for your part-time farm. Renting agreements should be in writing to save arguments.

If you have buildings, other options are calf rearing, breeding sheep, rearing pigs from weaners to pork or bacon weight, laying hens, broiler chickens, ducks, geese, turkeys and even alpacas and some crop growing.

When my business grew, I was still teaching full time, and it was too much for me to do both jobs. My parents worked for over a year with no wages, and then I employed part-time labourers to help out.

Full-Time Farming

You will need experience, either working on a farm plus a college course or some part-time farming of your own.

Start as a part-time farmer like me and keep developing your business, choosing one or more of the livestock enterprises with perhaps some arable, and eventually become a full-time farmer. I have achieved it, so there is room for others to follow me.

Eggs at Oak Tree Farm, ready for sale.

To obtain your land, you can start with renting one field – and then another and another, or you can rent a whole farm or possibly a council holding, or if you have enough money you can buy a field, fields or a farm.

Final Thoughts

A few people in this world will be left a fortune by relations or win the lottery, or they have had a very highly paid job, which would include working in television and films, or they may have created a very successful business.

Certainly television presenters and millionaire (even billionaire) businessmen have bought farms, as they think it is a better investment than putting money in the bank. Inheritance tax will not be incurred on farm land when they die, saving their children millions of pounds.

If you have millions to spend, you could buy a large country estate with the mansion and spend part of your life riding around in your Range Rover to make sure your labourers are working hard.

The icing on the cake in my farming world was to have met and married Sarah and to have fathered our two wonderful children, Jonathan and Roseanna.

Farming is hard work and in all weathers. A neighbouring dairy farmer has always worked hard, and when his son left school he joined his father on the farm. They worked 12 hours a day, 7 days a week, and after 6 years his son asked for a Sunday afternoon off work. His father's reply was, 'Why, are you not happy in your work?'

Whatever you are, be that
Whatever you say, be true
Straightforwardly act
Be honest in fact
Be nobody else but you.

In my opinion, you can call yourself a farmer when you do or have done most of the following:

1. Your border collie dog rides in the Land Rover more than your wife does.
2. You can remember the acreages of every one of your fields, yields of milk per cow, and fertiliser rates that you spread on the wheat and barley – but you haven't had time to remember your wedding anniversary or your wife's birthday.
3. You buy all your shirts and trousers from a country store such as Countrywide Farmers.
4. You certainly organise dates for your wedding, births of your children, and hospital visits so they don't clash with lambing and harvesting.
5. You and your family must not speak when the weather forecast is on the television or the radio.
6. You can tell if you are doing well by looking at the state of your farm and not the state of your house.
7. You paid more money for your tractor than for your motor car.
8. You have driven your car or Land Rover off the road onto the grass verge when looking over the hedge at your neighbour's livestock or crops, and almost crashed.
9. You don't have time to go on holiday.
10. When buying your wife's Christmas present, you hope the money can come out of the farm's account for tax purposes, and hope you can get the VAT back, and so you make sure you keep the receipt.
11. Your main topic of conversation is farming, even when eating your Christmas dinner.
12. You certainly haven't got time for all the paperwork!

Combining wheat into the sunset at Oak Tree Farm.

Our son, Jonathan, aged 8 – a second-generation farmer – judging sheep.

Our daughter, Roseanna – another second-generation farmer – showing her ducks. How wonderful for both Jonathan and Roseanna.

I have seen other farmers who have inherited land sell fields off. They kept chipping away, selling one field here and another field there – that's when the rot sets in – they end up with nothing. Never sell if you have worked hard like me to get it. Land is now the life blood of this family – land is a person's very own soul. Farming is the best job in the world!

This farm has provided me, and now my family, with a happy rural life, a wonderful place to live and a financial return – nothing could be better.

Here is some more advice – remember the following:

1. *More dogs than hogs*
   More women than men
   You will never get rich
   'Til God knows when.

2. *Never catch a falling knife or a running wife.*

3. *If you want to be remembered – owe some money.*
   If you want to be forgotten – lend some!

4. *Oh don't the days seem dark and long*
   When all goes right and nothing goes wrong
   And isn't your life extremely flat
   With nothing whatever to grumble at!

And finally:

5. *It's easy enough to be pleasant when the world rolls along like a song*
   But a man's worthwhile if he can smile when everything is going wrong.

The family at Oak Tree Farm.

Index

Pages in *italic* refer specifically to illustrations.

C

E

F

P

U

V

W

Y

Z